Shaping the Future

1 Making Physics Connect

Edited by Peter Campbell

Institute of Physics Publishing

Bristol and Philadelphia

Contributors

British Library Cataloguing-in-Publication Data

A catalogue record for this book is available from the British Library.

ISBN 0 7503 0621 1

Published by Institute of Physics Publishing, wholly owned by The Institute of Physics, London.

Institute of Physics Publishing
Dirac House
Temple Back
Bristol
BS1 6BE, UK

US Office: Institute of Physics Publishing
The Public Ledger Building, Suite 1035
150 South Independence Mall West
Philadelphia
PA 19106, USA

Printed by (to come)

Acknowledgements
Photograph on p.11 reproduced courtesy of The Royal Institution, London/Bridgeman Art Library, London/New York
Photograph on p.34 reproduced courtesy of The Cavendish Laboratory, University of Cambridge.
Poems on pp. 22 and 23 reproduced by permission of Faber & Faber, Oxford University Press, Jonathan Cape, Harcourt Brace & Company.

Martin Bazley is Assistant Education Manager at the Science Museum, London

Lyn Branson teaches at Bridgwater College, Somerset

Philip Britton is Head of Physics at Leeds Grammar School

Peter Campbell teaches at Highbury Fields Girls' School, Islington, London

Sarah Cramoysan teaches at Easingwold School, North Yorkshire

Lawrence Herklots teaches at King Edward VI School, Southampton

Jon Ogborn is Director of the Institute of Physics Post-16 Initiative

Becky Parker is Head of Science at Simon Langton Girls' School, Canterbury

Ben Rudden recently retired as Head of the Division of Applied Physics and Optoelectronics, School of Engineering, Northumbria University; he is now a QAA Assessor

David Sang is a freelance science writer and editor

Clive Sutton is chairman of the Education Section of the British Society for the History of Science

Elizabeth Swinbank is in the Science Education Group at the University of York, where she directs the Salters Horners Advanced Physics Project

with illustrations by *Ralph Edney*

Contents

Introduction

Peter Campbell

Scientific ideas have affected the way we all experience the world. Philosophers, poets and novelists have been touched by them. But conventional science teaching does us all a disservice, misrepresenting the nature of science and at the same time alienating learners. There is a great need to re-establish the human-ness of science.

When I started my teaching career, I thought physics had to be taught in a linear way. Conditioned by my own physics education, I tried first to establish basic principles and then gradually build the edifice that is physics, using a combination of reasoning and experiment to convince the students. But twenty years of classroom teaching have consistently shown me this is not how learning takes place. The page entitled *What Students Say* illustrates typical student views.

Physics is unlike music or art, which feel holistic even when we analyse their parts. When analysing a painting, for example, we are always aware of the whole image. Naturally we teach the content of a physics syllabus by parts, but we must also take care that students do not get a false image of the whole – the whole subject, its nature, its part in human culture, and its relationship to social and political life. Without such connections the abstract method of physics loses its purpose, and understanding is cut adrift.

Points of reference used by teachers to orientate students should at the same time be motivating. This, in part, is why the science–technology–society connections need to be made. Not that students are familiar in detail with many technologies or understand how they are driven. But often they want to know, because they need to know to grow.

The same is true of the natural world: understanding our place in Nature, our place on Earth and in the Universe, is a deep-seated need. Exploiting these motivations can help students grasp physics concepts.

Unfortunately, students can succeed at GCSE level secure in the false notion that science deals in facts, in discovery of truths, rather than appreciating that it always bears the birthmarks of its human origins.

Knowing about the historical context from which new ideas were invented sheds light on the ideas themselves, as Clive Sutton, in *What are Physics Lessons For?* and Jon Ogborn, in *Michael Faraday Making Science* demonstrate. David Sang's *Sketch Notes* tie these ideas more closely to the National Curriculum and practical issues for classroom teachers. We should try to convey physics as a way of seeing the world.

In this sense, learning physics is just the same as learning about the visual or performing arts, or learning about literature. In every case, there is a personal vision which interprets a common, objective reality – or refers to an earlier cultural artefact. Robert Frost (in his poem *A Tuft of Flowers*) has a character say: "Men work together ... whether they work together or apart."

Science fairly clearly is a collective enterprise where theorists, experimentalists, technicians and managers all play their parts. (See Emilio Segre, *From X-rays to Quarks*, for a wonderful annotated diagram showing the interrelationships between parts which make the body of physics.) Frost of course meant to portray the interdependence of all humanity. The everyday world, including the mass media, largely ignores this vital insight, emphasizing instead 'great men' and 'critical events'. It is very important that anything which calls itself education draws attention to culture broadly, as both process and product, humanity's best collective enterprise.

What are the opportunities for presenting an international dimension in physics education? Of course, there are finance-driven 'big physics' collaborations like CERN. Becky Parker writes enthusiastically about her class visits, in *CERN and other wild trips*.

Visits to other industry or research institutions may serve the same purpose. We can also help students to get beyond the disturbing images of the Third World they encounter through mass media. Appropriate technologies, such as pedal-driven machinery for example, can be explained with basic theory – and this then gives a purpose for understanding mechanics. It simultaneously recognises and respects the way people make use of science principles in designing technologies appropriate to their limited circumstances.

Martin Bazley, in *Using the Internet* discusses how it can be used to develop independent learning skills while students travel the virtual world and broaden their horizons.

Many teachers, in classrooms around the country, connect physics to music, religious education, politics, visual arts, etc. My *Poetry in the Physics Classroom* shows just one way of bringing together science and the humanities.

What kind of people are physicists? Do they experience everyday life and emotions, or are they social misfits concerned only for ideas? The *Famous Physicists* exercise provides a way to start exploring traditions and conventions in science, gender issues and some of the drama of real life. These can be explored further using biographical and autobiographical writing, plentiful in twentieth century physics. I hope extracts from such writing will begin to find their way into learning resources.

Until they begin to draw conclusions, students aged 14–19 are continually asking themselves: "What kinds of scientists are there?", "What contexts do they work in?" and "Could I (and shall I) become a scientist?"

What images do students form while in our classrooms? It is vital that physics educators help them answer these kinds of questions. It is at this personal level that the gender and multicultural dimensions are most powerful. Fortunately there is already a wealth of literature on the important issues of gender, multiculturalism and anti-racism in science education. Here Lyn Branson gives a strongly personal view in *Women in Physics*. Lawrence Herklots writes about challenging students positively in *Presenting Physics*, suggesting activities which could be used with any course.

We then have four contributions which demonstrate aspects of existing courses which foster understandings of how science develops and how scientists work. These also encourage student self-confidence in the subject. Lyn Branson describes how physics is taught and assessed in *GNVQ Advanced Science*. Sarah Cramoyson and Philip Britton describe positive learning effects of the *NEAB Option 'Turning Points in Physics'*. Lawrence Herklots describes the benefits of *Nuffield A level Physics Research and Analysis*. Finally, Elizabeth Swinbank explains how *Salters Horners Advanced Physics* makes connections and motivates learners.

We include a brief look at undergraduate education. Ben Rudden, in *Problems and Challenges in Higher Education*, discusses how the image of physics impacts at that stage. Finally, there is an annotated bibliography.

I hope you enjoy dipping into the booklet, finding points to agree with and also those where you differ. Our more general aim is to stimulate discussion.

What Students Say

Students who have not chosen physics at A level:

" With my GCSE results, I could have done physics – but I am more interested in people and ideas."

"Everyone knows physics is a difficult subject, which includes demanding maths. Although I am interested in some topics in physics, I thought my grades would be poor so I didn't do it."

"I was really interested in physics at GCSE, but I didn't think it would be useful for my future. I don't want to become an engineer."

Students who have just completed an A level physics course:

"Some of the ideas studied in this course were potentially very interesting. Things studied in class can be applied to the world around us. I feel this isn't emphasised enough in the syllabus."

"The teacher must motivate the students. There weren't as many discussions and debates as I expected."

"I especially enjoyed my teacher bringing in stories and relating them to the physical world. This taught me that physics is out there – everyday."

Physics undergraduates' comments about their A level course:

"Too little finding out for myself left me inadequately prepared for university. Lively discussion in class or extra-curricular was as beneficial as taking notes from lectures."

"Humanise the subject with greater mention of the inventors and great physicists of our time."

"Some part of the course should include current physics such as the Hubble Space Telescope or the exploration of Mars. This would enthuse students and make them feel that physics is something exciting, happening now, and moving in leaps and bounds."

"Physics teachers must make physics fun because it is very easy for physics to be considered hard and boring, detached from the real world. Our course was geared to studying for marks and not for interest."

"People generally choose physics because it is needed for their future, not because of the weird and wonderful parts."

Conclusions from the Institute of Physics student survey (1992):

"Among students not studying physics A/AS level, 58 per cent think their teachers had not made physics interesting at GCSE level"

"This and other studies suggest that, at any level, the greatest influence causing students to study physics is the interest developed in the subject at the previous level"

"The role of the study of physics as a general intellectual training on a par with the classics or humanities is not generally recognised by A/AS level students."

"Careers material should be developed which shows... young physics graduates successful in non-technical careers (e.g. banks, accountancy) so that such people will be scientifically literate and sympathetic to physics as a discipline."

1 What are Physics Lessons For?

Clive Sutton

... to learn about Nature? ... or to learn about human thought?

Question 1. Where on earth is the humanity in a circuit board?

This question was put to me recently by an irritated parent who was grumbling about what she saw as the aridity of science lessons for her teenagers. She evidently sensed that when most of the effort goes on learning 'how a circuit works', both teachers and pupils lose sight of the adventure of ideas that lies behind such knowledge. They stop attending to the human struggle to understand what might be going on.

I know what she meant, but her question might put any of us on the defensive because it seems so out of line with our intentions and beliefs. Science is not arid to us; we constantly try to communicate its excitement and significance and we succeed in drawing in many youngsters to share the excitement of it. Despite our best efforts, however, others retain alarmingly stereotypical perceptions of physics:

- boys' toys!
- things, not people!
- a nerdish pre-occupation with facts!
- exclusion of one's own thoughts and opinions! (see further reading - Lemke)
- large servings of impersonal and very serious knowledge to be accepted rather than discussed!

We need to understand how such stereotypes arise, for, quite apart from alienating some learners, they project a totally inaccurate picture of science and scientists, at variance with what we want to convey. To let them persist would be to go on allowing a public to grow up with basic confusions about the nature of science.

Question 2. If citizens' images are confused, what part do school lessons play in causing that confusion?

As I see it, all the dangerous confusions are related to differences between science as an activity and science as a product (the body of accepted knowledge which that activity has generated).

They also involve learners' and teachers' beliefs about what the lessons are mainly for – to learn the conclusions?... or to understand the adventure? Do pupils come to learn about Nature as we now understand it, or do they come to learn about how we got to this point?

The tension between process and product has been discussed rather inconclusively for decades, and should perhaps be recast in different words now. Drawing on the writing of Bruno Latour, we might now call it a tension between 'science-in-the-making' – new science, still debatable

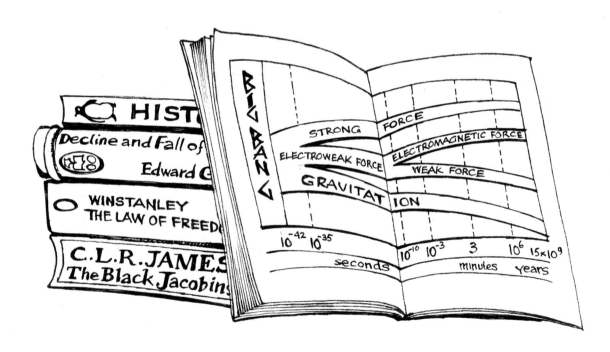

- **Despite our best efforts, some people retain alarmingly stereotypical perceptions of physics**

- **Too many learners hear the facts as plain knowledge about the natural world which needs no debate**

science, fresh science – and 'ready-made science' – old science, established science, accepted science – taken for granted and no longer debated and therefore to some extent stale science. In the latter, the known facts are the dominant feature, whereas the former is all about ideas and evidence.

I should elaborate that last point a little. Practising scientists at the growth-points of new thought live in a mental world of ideas and evidence which are the subject of constant debate and test. Every discussion of an experiment is concerned with what will count as good evidence and how someone's ideas can be adequately tested. There is no doubt at all that this 'science-in-the-making' is an activity of human beings, involving previous *ideas*, judgement, opinion, argument and debate.

When school science does not manage to convey those features it is missing the essence of science. That is a tragedy, I think. And doubly so when teachers see themselves as trying to show this very point, but for the learners it gets lost in the mass of accepted facts which are now so taken for granted that they don't even sound like human ideas any longer. Too many learners hear the facts as plain knowledge about the natural world, which needs no debate.

On the other hand, for any learners or parents who have some sensitivity to personal engagement in learning, the experience of hearing all these ready made facts is arid and excluding. This need not happen if we could have a better negotiation of what the science lessons are for in relation to the two facets of science.

Question 3. What do we encourage learners to expect?

If we are not to fall victim to established stereotypes and ill-informed fashion I think it is necessary to negotiate more carefully about what they can expect and what we can hope to offer. And why.

A science lesson might be a time to learn about nature, i.e. how the natural world works. On the other hand, it could be a chance to learn about what people have thought and said about nature, to learn about how we think the world

works, what the evidence is, and what the arguments are.

Simply to pose the contrast clearly might be enough to prevent too heavy an emphasis one way, but of course it is the latter which seems to me to be a far more engaging purpose, and one which is much more faithful to science-as-practised. It can also make the scientists' contribution to our culture intelligible and persuasive, without rejecting other peoples' contributions.

(Note the warning in early versions of the National Curriculum about over-confident scientific arrogance that places science as the only valid form of knowledge: "Pupils should begin to recognise that while science is an important way of thinking about experience it is not the only way." Historians, artists, even theologians, are part of the human striving to understand, and scientists are a part of this striving too, with a marked record of 'success', or mainly success, as we now see it.)

Question 4. What is practical work for?

Does a circuit board give direct experience of 'what happens' so that a person can learn the science substantially by doing the practical work? My view is that it only makes human sense when we know how and why someone designed the board to help us explore the ideas of Coulomb, Ampère, and Faraday, which in turn depended on ideas articulated by Hauksbee, Gray, Franklin and others back in the eighteenth century.

I think that pupils will only get a sense of involvement in the total project ('Humanity understanding the world around us') if they can 'hear' the voices of thinking scientists sorting out what seems to be going on.

The actual words matter, as when Stephen Gray wrote in the early eighteenth century about the first experiments on what we now call 'electrical conduction': "I was much surprised and concluded that there was certainly an attractive virtue communicated to the cork by the excited tube." And again later when he wrote of how this 'electrical virtue' could somehow be 'communicated' along threads that were

- **The recovery of the human voice in science seems to to be an urgent problem for schools, and for our culture generally**

- **Science lessons should be the study of what people have said and thought about Nature**

hundreds of feet in length. The voice of the thinking scientist, sorting out what seems to be going on, seems to me to be a vital ingredient.

Indeed the recovery of that human voice in science seems to me to be a very urgent problem for schools, and for our culture generally. I wrote about it as follows in 1992.

> *.... stop thinking of science lessons as the study of Nature. Science itself may be a study of Nature, but science lessons should be the study of what people have said and thought about Nature. The main object of interpretive activity should be not the circuit itself, but what someone has said about the circuit... This person, who told the 'story' we are considering: what was he trying to say?*

Science lessons should be the study of systems of meaning which human beings have built up. Practical work is necessary in order to get a feel for those systems, and to give an understanding of what the evidence is which supports the scientific view, but it should not be thought of as the source that ideas come from.

Further reading

1. To hear spontaneous anger about feelings of exclusion in science lessons, read chapter 5 of Jay Lemke, *Talking Science*, Ablex Publishing Corporation, New Jersey, 1990.

2. To examine a fuller argument about why science lessons should be based on the voices of scientists, and how it can be done, see chapter 9 in Clive Sutton, *Words, Science and Learning*, Open University Press, 1992. I have also extended the argument in *Recovering the human voice in science: an urgent problem in school*, which is now available on the World Wide Web at *http://indigo.stile.le.ac.uk/~edu/STILE/Staff/voice.html*

3. For early language about electrical phenomena it is still useful to go back to J.B. Conant and L.K. Nash (editors), *Harvard Case Studies in Experimental Science, Vol II*, Harvard University Press, 1957. For recent material on a different topic which embeds scientific thought in a wider exploration of human ideas, try chapters 2 and 3 of Arthur Zajonc, *Catching the Light: The Entwined History of Light and Mind*, Oxford University Press, 1995.

DISCUSSION POINTS

- Can we, as Clive Sutton asks, talk effectively about 'what the arguments are'? Discoveries involve not only thinking of the right answer, but of disposing of a multitude of wrong answers. The danger is that we tell false history, with everything more clear-cut than it was.

- Does Clive Sutton go too far? Physicists study Nature, not themselves. They count what is done as valuable when a clear and as far as possible indefeasible account of Nature emerges. Certainly how it is done matters, but what the study of Nature produces also counts for a lot.

- Is 'ready-made science' really stale science? It is live enough when there to be put to use. The production of spectra is ready-made, but using the ideas and techniques to find quasars has plenty of life in it.

2 Michael Faraday Making Science

Jon Ogborn

Michael Faraday (1791–1867)
1851 (litho)
by Thomas Herber Maguire
(1821–95) (after)

Michael Faraday is simple and clear. He writes personally, at once meticulous and boldly imaginative.

Discussing making science with students

Among makers of science in the past who can still speak to us today in their own authentic voices, Michael Faraday surely ranks very high. I have chosen four episodes from his writings. Each offers material for a serious and challenging discussion of how science ought to be done, and puts into question certain commonly held ideas about the nature of scientific work.

The reprinted volumes of Faraday's *Experimental Researches* are an invaluable resource. I treasure my copy as you will treasure yours.

Faraday jumps to conclusions?

Here is Faraday in 1833 in the middle of investigating how liquids can be decomposed by an electric current:

"I was working with ice ... when I was suddenly stopped in my progress by finding that ... as soon as a thin film of it was interposed, in the circuit of a very powerful voltaic battery, the transmission of electricity was prevented, and all decomposition ceased.

At first the experiments were made with common ice, during the cold freezing weather of the latter end of January 1833; but the results were fallacious ..."

Experimental Researches, paragraphs 380-383

Faraday describes a more careful design of apparatus and then tells us how "not the slightest deflection of the galvanometer occurred" while the water remained frozen, but how there was a current when the ice was thawed. He tried the experiments many times in many ways: "It seemed at first as if occasional departures from these effects occurred, but they could always be traced to some interfering circumstances." For example a film of fluid on the surface of the ice.

So far, we see just the careful meticulous work of checking and cross-checking a result. But Faraday immediately makes an imaginative leap:

"As it did not seem likely that this law of the assumption of conducting power during liquefaction, and the loss of it during congelation, would be peculiar to water, I immediately proceeded to ascertain its influence in other cases, and found it to be very general ... On fusing a little chloride of lead by a spirit lamp on a fragment of a Florence flask, and introducing two platina wires connected with the poles of the battery, there was instantly powerful action ... on removing the lamp, the instant the chloride solidified all current and consequent effects ceased."

Experimental Researches paragraphs 394-395

We all know that one swallow does not makes a summer; that instances do not make a law. Yet here we have Faraday jumping from the detailed particulars of one experiment to a general principle ("this law") even before it was checked in a second case. What is he up to?

To understand, you need to know more about Faraday's religious beliefs. He belonged to a minute sect call the Sandemanians, which broke away from the Scottish Presbyterian Church. Their central belief was that the sole duty of people is to live according to a plain and literal understanding of the Bible, an understanding to be reached in common with other members by simple, honest and careful reading of the Word of God.

General rules for right living had to be worked out from particular passages of Scripture. It may now be a little easier to understand Faraday's belief that careful observation of particulars, meticulously checked, can yield much more than particulars. They suggest to him the ground plan of part of the Universe. Is doing science just a matter of making an honest reading of the 'Book of Nature'? The idea merits thoughtful discussion.

- Few have been more conscious of the power of words than Faraday

- Faraday wants to make it 'natural' to think in his way and harder to think in the older way

Faraday makes up words to suit himself?

Few have been more conscious of the power of words than Faraday. He created the words we now use to describe electric currents passing through solutions. In 1834, the phrase, 'electric currents passing through solutions', would not have been used. People spoke instead of 'poles' dipping into the solution, and of effects 'on' the solution near the poles. There was no idea of anything going through the solution.

Faraday had a different vision. So he set about inventing words whose purpose was to act to enforce that vision:

> *"The theory which I believe to be a true expression of the facts of electro-chemical deposition, and which I have therefore detailed in a former part of these Researches, is so much at variance with those formerly advanced, that I find the greatest difficulty in stating results, as I think, correctly, whilst limited to the use of terms which are current with a certain accepted meaning."*
>
> *Experimental Researches, paragraph 661*

Here is the view that Faraday was concerned to dispute:

> *"Grotthuss, in the year 1805, wrote expressly on the decomposition of liquids by voltaic electricity. He considers the pile as an electric magnet, i.e. as an attractive and repulsive agent; the poles having attractive and repelling powers."*
>
> *Experimental Researches, paragraph 481*

Faraday wants us to think, not of attractions and repulsions, but of things moving through the liquid. He commends his friend Auguste de la Rive for suggesting that:

> *"The current from the positive pole combining with the hydrogen ... carries the substances it is united with across to the negative pole ..."*
>
> *Experimental Researches, paragraph 489*

The key words are 'current' and 'carries'. By 1834, Faraday was ready to try to change the world by creating new words. Having "deliberately considered the matter with two friends" (one of whom was William Whewell, who later coined the term 'scientist'), Faraday says, " I propose henceforward using certain other terms":

> *"The poles, as they are usually called, are only the doors or ways by which the electric current passes into and out of the decomposing body ... In place of the term pole I propose using that of electrode ..."*
>
> *Experimental Researches, paragraph 662.*

A footnote gives the source of the term 'electrode': ηλεκτρου and οδοσ (a way). Faraday also coined the term 'ion' (traveller) for the moving electrical particles. He writes: "I am fully aware that names are one thing and science another." But his purpose is clear. He wants to make it 'natural' to think in his way and harder to think in the older way. It worked: today we see electricity his way.

The moral? It is that scientific terms are often a matter of power, not a matter of precise distinctions. Scientific words are chosen to 'make you see it my way'; as indeed they are too in everyday life.

Faraday gets the right answer from a mistake?

Nobody is in favour of sloppy thinking, and Faraday was no exception. But it remains the case that one of the most important new ideas ever introduced to science was introduced by Faraday on the basis of a rather simple mistake.

Faraday was the first to dare to propose that magnetic fields should be thought of as real entities, not as mathematical fictions or calculating devices. His idea was that the magnetic field should be thought of as materially present in the space around a magnet or coil. Here is his remarkable reason for thinking so:

> *"It appears to me, that the outer forces at the poles can only have relation to one another by curved lines of force through the surrounding space; and I cannot conceive curved lines of force without the conditions of physical existence in that intermediate space."*
>
> *Experimental Researches, paragraph 3258*

- One of the most important new ideas ever introduced to science was introduced on the basis of a rather simple mistake

- Anyone who supposes that Faraday would allow contrary facts always to rule out a theory is wrong

The idea of curvature being special to magnetic fields is just a mistake. Despite his experimental gifts and imaginative power, Faraday was no mathematician, and understood nothing of vector superposition. He thought that gravitational and electrical 'lines of force' were necessarily straight, not knowing that the resultant field of a number of sources can be represented as curved lines.

But, however wrong its grounds, Faraday's vision was essentially right. He rejected a mechanistic universe made purely of particles in motion in empty nothingness. His essentially religious belief in the 'powers' of God being made manifest in Nature led him to look for such powers as real constituents of the world. Nowadays we regard fields as a basic part of the description of Nature.

Furthermore, this insistent imaginative vision led Faraday to one of his greatest speculative ideas; that light might somehow be electromagnetic. This really was 'making the field real':

> *"The view which I am so bold as to put forth considers, therefore, radiation as a high species of vibration in the lines of force which are known to connect particles and also masses of matter together ... The occurrence of a change at one end of a line of force easily suggests a consequent change at the other. The propagation of light, and therefore probably of all radiant action, occupies time; and, that a vibration of the line of force should account for the phaenomena of radiation, it is necessary that such vibration should occupy time also."*
>
> *Experimental Researches: Thoughts on Ray-Vibrations 1846*

It took James Clerk Maxwell to turn this idea into more than speculation, but the seeds lay in Faraday's imagination. The argument to get there was wrong, but the answer was right enough to sow the seeds of much of our present understanding.

Faraday won't give up even in face of the facts?
Faraday was devoted to facts; to simple clear empirical evidence. But this is not the whole truth. Believing as he did

in the importance of the forces of Nature, and in their essential unity – a belief wholly consistent with his religious convictions – Faraday constantly sought relations between them.

In the case of electricity and magnetism he was stunningly successful, discovering electromagnetic induction. Following the same belief, he was sure that there must be a connection between the other great known force, gravity, and electricity and magnetism:

> *"The long and constant persuasion that all the forces of nature are mutually dependent, having one common origin, or rather being different manifestations of one fundamental power, has made me often think upon the possibility of establishing, by experiment, a connexion between gravity and electricity ..."*
>
> *Experimental Researches, paragraph 2702*

His approach was experimental. Faraday describes a long series of experiments, involving dropping coils under gravity, and making masses oscillate inside coils. Occasionally there seemed to be positive results, but always when tracked down they turn out to be 'fallacious'. In the end he found nothing.

We know what Faraday 'should have done' in face of this disagreement between theory and experiment: give up the theory. Instead, he ends his account of his many failed experiments with the following stunning remark:

> *"Here end my trials for the present. The results are negative. They do not shake my strong feeling of the existence of a relation between gravity and electricity, though they give no proof that such a relation exists."*
>
> *Experimental Researches, paragraph 2717*

Anyone who supposes that Faraday – the arch empiricist – would allow contrary facts always to rule out a theory, is wrong. Faraday knew where he wanted to look, and indeed people are still looking in the same corner today, now under the name of Grand Unified Theories.

Full success eludes them too. So far.

References
Michael Faraday, *Experimental Researches in Electricity*, Dover Edition (three volumes bound as two), 1965.
Geoffrey Cantor, *Michael Faraday, Sandemanian and Scientist.*, Macmillan 1991.

Footnote
The Sandemanians were a minute religious sect, numbering only a few hundred in Faraday's time, which had broken away initially under the leadership of a Scot, John Glas, from the Scottish Presbyterian Church. Then, and under the later leadership of Robert Sandeman, the sect opposed all worldly involvement, holding that the sole duty of people is to live according to a plain and literal understanding of the Bible. Because that understanding was held to be simple and direct, with no room for disagreement, the sect insisted on complete agreement of all its members. Those who could not accept the common view were excluded, 'put away'.

DISCUSSION POINTS

- The Faraday example is probably timeless. But we have to be careful that other examples don't date, or to replace them when they do. The 'discovery' of 'delta rays' once made an interesting tale, but is now forgotten.

- Is there a need for modern examples similar to that of Faraday? Instead of Faraday coining the term 'ion' what about Gell Mann and others coining terms like 'quark', 'strangeness' and 'charm'?

- Is there a danger that Faraday is a very special case? Earlier scientists are often unintelligible to students because the problems they deal with look so peculiar; ideas and language have changed. Later scientists are often unintelligible because of the importance to them of mathematical physics .

3 Sketch Notes: Bringing Some History into School Physics Teaching

David Sang

Rationale

- Occasionally a story from history is a useful way to introduce an idea. Historically, a common-sense but incorrect idea may have been replaced by a new idea. Pupils may have to go through the same process.

  A geocentric world view is common sense, but we are taught that it is incorrect long before we are introduced to the evidence which supports the heliocentric view. Pupils are often taught about the solar system in terms of the work of Kepler, Copernicus et al.

- A story from the history of physics may illustrate some of the ways in which scientists work (Sc1).

  The story of Becquerel and radioactivity is not one of a chance discovery, as is sometimes suggested in textbooks. His experimentation was clearly directed by the ideas he had, and he developed his ideas further in the light of his experimental findings.

- Examples from the history of physics show that scientific ideas change. Hence our present scientific ideas are only temporary and will change (Sc0).

  Pupils are often introduced to the idea of indivisible atoms, the plum-pudding model and the nuclear model of atoms. Advanced students may be brought slightly more up-to-date with the quark model of hadrons.

- A story from the history of physics may show how scientific ideas develop in a social context and reflect that context (Sc0 again).

  The military applications of physics can provide many instances – for example, the rapid development in understanding of fission during World War 2.

- History can provide links between disparate aspects of science, bringing the subject together.

  An understanding of radioactive decay resolved Victorian debates about the age of the Earth, and revealed the source of energy which drives the motion of tectonic plates. This is an area which unites atomic and thermal physics with geology.

James Clerk Maxwell
Poet of Waves

Cautions

- Teachers need to be convinced that this is an important aspect of the science curriculum. OFSTED are looking for Sc0 content in schools' schemes of work.

- Teachers don't feel confident – so it can help to produce self-contained materials.

- It takes a team of people to produce new materials – practising teachers, writers, editors, experts on the history. These roles may overlap to some extent in particular individuals, but most of us cannot cover all aspects.

- There is a lack of time for teaching – so it is important that historical materials are compact and aren't seen as an extra. Science content can sometimes be taught through history. Perhaps a short historical unit will serve as a homework activity, or can be used when the teacher is away. It might be only ten minutes' worth.

- Information needs to be reliable. A lot of recycling goes on in published materials. We need to publish new, reliable information which may become incorporated in the canon of received wisdom.

- Students are unlikely to engage with materials which involve only reading. Activities need to be designed so that students interact with the text.

- History needs different sorts of activities – discussion, consideration of sources, role play, etc. – of which teachers have less experience. Once they have tried one of these different approaches, they often report that the experience wasn't as bad as the anticipation.

DISCUSSION POINTS

- Remember also that experiments have a history. The simple work we do with thermometers and heating water goes back all the way to Joseph Black.

- Certainly teachers lack knowledge and confidence in tackling historical questions. But as well as paper resources they also have human resources at hand; some collaboration with history and English teachers could be profitable.

- If syllabuses are to include some material making physics connect, then they need to provide support material, just as they have to when new topics (e.g. particle physics) are introduced.

4 CERN and Other Wild Trips

Becky Parker

"Why are you thinking of doing A level physics?"
"I like physics and I want to go on the CERN trip."

Two years later, at the end of the course:

"Are you glad you did A level physics?"
"Yes, the CERN trip was the best time ever."

This may seem a bit far fetched but is a real conversation with one of my students who was actually a dancer and decided to do physics for rather, you may say, dubious reasons!

Our visit to CERN, which is a six-day, five-night trip has become one of *the* school trips, I think because of its awe-inspiring nature and the great fun we have. The students find the visit to the accelerator and detector mind blowing and there is nothing to compare to it. No video gives the sense of the extreme vastness and enormity of it all – even more scary to think that this biggest apparatus in the world searches for the smallest of particles in the universe.

I recently went to look at DESY in Hamburg to see if it could be an alternative to visiting CERN, but while DESY is great, it is the scale of CERN that opens the students' minds to what a major European and international collaboration means in practice.

They don't necessarily understand all the physics – that would be left for them at University! It's being immersed in the language and seeing the way such an experiment runs, with all the masses of computer power and tonnes of spaghetti wiring, and the chatting over lunchtime in the CERN canteen where they see famous people.

"That man was on TV", they say pointing out John Ellis.
"Well that's because he's doing this amazing work HERE!!"

One of the high spots of our most recent visit, was Frank Close popping in to say hello and comparing the 'accelerator' JJ Thomson used to discover the electron – a model of which Close held in his hand – to the vast 27km of the LEP ring.

The thing is the students are right there and they are getting a feel of what a buzz physics research has. On our first trip, we were there at the discussions about whether the top quark existed or not. Each time it's: "What about the lepto quark?" or "Anyone seen the Higgs boson?"

They also get the mix between physics and fun for they hear about the physicists going off skiing and they see the beautiful Mont Blanc in the distance – the world there is really rather magical.

I often worry that we can portray physicists either as a bit in the clouds, having an ethereal countenance, or so committed that they never do anything but ponder on equations. It is lovely therefore to see people from CERN playing pool in the local centre.

There is a full social programme on the trip: we work hard and play hard!

The students love this because they know they will also have some good laughs whether it's in the bowling alley, on the air hockey tables or singing and dancing madly at the karaoke! Each to their own but you've got to make the whole trip a great time.

On the way back via Paris we visit *la Geode* and see a fantastic IMAX film as well as having a good look round the Science Centre in Paris, *La Villette*.

That's what I mean at the beginning about fun trips – we took our students to Alton Towers, and my latest plan is a day at a chocolate factory followed by a day at Alton Towers. Plenty of physics there! Better keep clear of the big rides!

Physics must be inspirational, interesting and enjoyable. Many students are really thrilled by going on a CERN trip. At £200 for six days, it is also not very expensive – and sometimes sponsorship helps to finance particular students.

It is obviously a big worry to run, but it is also inspiring for the team of staff – we all come back exhausted yet raring to go!

DISCUSSION POINTS

- Local visits are more important, giving insight into possible future careers for students – not many will work at CERN-like projects.

- Having a good tour guide makes all the difference. Robot-like commentaries are useless.

- It would be really helpful if physics teachers' networks helped with planning local visits.

5 Using the Internet

Martin Bazley

Although most people would agree that the Internet has enormous potential to enhance learning, it is sometimes difficult to distinguish current, real opportunities from the surrounding hype.

Use of the Internet in education is still in its infancy. Indeed some educators view this nascent medium's most enthusiastic promoters as *enfants terribles*. Before looking at some of the Internet-based projects and resources that may be of use in this context, therefore, it may be worth stepping back for a moment to consider the more general issues involved.

The first section of this article reviews the primary reasons students and teachers might have for using the Internet. The second section considers briefly some of the likely 'fringe benefits'. The third section goes on to highlight a few specific sites, projects and ways of using the Net that may enhance physics education.

Reasons why students and teachers might use the Internet

Clearly a learner might have one or many different motives for using the Internet. Here is one way of categorising the most likely ones:

To find information

This is the way in which most 'newbies' expect to use the Internet. The 'big library in the sky' is perhaps the most common metaphor, though not a very helpful one. With most libraries you have a right to expect certain standards in the reliability and provenance of the information, as well as the way in which it is organised.

There are repositories of information on the Net that meet these criteria, but plenty of others that do not. Teachers in particular are often very disappointed when they first gain access to the Web and use a generalised search engine or 'portal' (AltaVista, Excite etc.) to look for items of interest using one or two keywords. Most frequent Web users find their own 'specialist' sites and search engines, and will only use such blunt tools when looking outside their field – when booking a holiday, for example. Some useful sites to use as starting points for information searches are given below.

Another worryingly prevalent misconception is that giving students access to information means they will learn something. This fallacy seems especially beguiling in relation to the Internet, as there seems to be an implicit assumption that viewing something on a computer is somehow 'interactive' and that creation of 'knowledge' in students' brains is virtually automatic. Perhaps this point is too self-evident to make in this context, but clearly it is not enough simply to direct students to a website and ask them to 'make notes' or 'learn about wave particle duality'.

To access presented 'content'

It is perhaps more reasonable to expect meaningful learning when students go to a website specifically designed to encourage learning, i.e. where use of the website involves knowledge generation as a process, rather than simply absorbing a product.

Implicit in these resources will be different theories of learning, from behaviourist 'tutorial' style packages to more 'constructivist' approaches, both of which can be useful. There are examples of different types in the websites mentioned below.

To communicate/collaborate with others

The Internet is first and (should be) foremost a communications medium. Learning by collaboration across the Net has many advantages over accessing 'delivered' educational 'content', not least that it reinforces the view of science as activity rather than as 'finished' product. The learners often feel more involved and may have significant input into the scope of the collaboration. In sharing knowledge with those

- some educators view this nascent medium's most enthusiastic promoters as *enfants terribles*

- A worryingly prevalent misconception is that giving students access to information means they will learn something

in different cultures, education systems, across age ranges and so on, some of the humdrum elements of everyday communication may become of great interest.

A comparative project dealing with, say, access to clean water – something most UK students take for granted – could become a fascinating subject with great potential for meaningful learning, provided the project is well structured.

Possible 'fringe benefits' of using the Internet

The most obvious benefit is access to an enormous amount of information and educational 'content', some of it potentially extremely useful. The shortcomings of the Internet as a reference tool mentioned above also can be viewed as opportunities for students (and teachers) to develop information-handling skills.

For example one quickly develops awareness of different types of information sources, bringing up issues of reliability (which most teachers manage to avoid addressing in the context of 'text books'), bias, completeness, register, target audience and so on. In more general terms, selecting, prioritising and 'managing' the information they encounter is surely one of the most important skills students can acquire, and this aspect is often more 'in your face' when using the Internet than print media.

Using the Internet to communicate also involves developing important skills in new ways. At the mundane level there are the continuously evolving online conventions ('netiquette') that some students (boys?) seem to enjoy mastering to the exclusion of others.

There are also positive aspects to the strategies people employ to build up a picture of their online correspondents. They say that on the Net, everyone is equal. While ludicrous if interpreted at one level, it is often true that once meaningful communication is established it is easier to focus on the things that really matter.

Videoconferencing can be highly motivating and a good way to initiate collaborative projects, but it is not a good medium (given current setups) for considered discourse. Far

better for this purpose is email, where the participant has time to think about what they are 'sending' rather than simply reacting instantaneously. Similarly live 'chat' channels, while initially more motivating, have not proved very useful learning situations.

Returning to the misleading 'Internet as library' metaphor, perhaps the main disadvantage of this uninformed view of the Net is the way in which it reinforces, at a very fundamental level, the 'delivery' model of education. It is my perception that this concern is at the core of the disquiet expressed by those opposed to an increase in the use of ICT in education.

If educators can learn to exploit the full potential of the Internet as a communication medium, however, ICT will help us to move away from the one-way transmission of information towards more interactive ways of learning.

In common with many educators today I share the 'constructivist' view that it is not possible to 'transmit knowledge', since construction of knowledge is an active process that can only take place in the mind of the learner. At issue here is whether use of the Internet can help to stimulate that process for those learning physics, and whether it can also support the *Making Physics Connect* aim of broadening the way physics is perceived by students, teachers and others.

Sites

Mentioned here are some sites that may be useful in the context of physics education. Roughly speaking they move from 'information retrieval' sites, through 'presented content' to more 'collaborative' projects. This is obviously a highly selective sample and in any case new sites are springing up all the time. For the latest list please check the *Institute of Physics* website.

The *Institute of Physics* and *ASE* websites are of course excellent starting points for physics and science orientated information searches.

ScienceNet, the online counterpart to the very popular ScienceLine telephone enquiry service, has a good database

- It is not enough simply to direct students to a website and ask them to 'make notes' or 'learn about wave particle duality'

- Selecting, prioritising and 'managing' information is one of the most important skills students can acquire

of questions and answers and also allows enquirers to 'ask an expert' by email.

The *SORT* (Schools Online Resources for Teachers) database is self-explanatory.

The *Southeastern Michigan Math-Science Learning Coalition* site includes a substantial database of 'physical science lessons'.

The *National Grid for Learning's* Virtual Teachers' Centre also has resources which may be of interest.

Working in a museum I have to mention the highly comprehensive *Virtual WWW Library of Museums* website, as the use of museums as 'mainstream' educational resource centres seems set to increase over the next few years. The site may lead you, for example, to the *Science Museum* website, with news of the forthcoming galleries Making the Modern World ('a cultural history of industrialisation'), and the Wellcome Wing ('a new wing to celebrate the cutting edge of contemporary science in the new millennium') both rich in stories to help contextualise the application of physics. There are also online exhibitions including the Institute of Physics-supported *Life, the Universe and the Electron* and of course access to information about the Classical and Modern Physics collections.

A reliable source of some excellent 'presented' content is the international *Science Learning Network*. Already in their growing database of online resources are *Science of Cycling*, *Science of Hockey*, *Auroras: Paintings in the Sky*, *Scanning Electron Microscope* and one called *Exploring Leonardo*, which takes a very human look at the polymath's contribution.

Head of Science, *Vernon Levy's* prize-winning website, contains some cleverly constructed 'dynamic science activities' on topics such as mechanics, refraction, radioactivity, and day length.

The US-based *IPPEX* site carries some 'interactive physics modules' along with a 'virtual tokamak' and other fusion-related activities.

Moving to projects where the collaborative element is more explicit, there is the ASE-supported *Science Across the World* (SAW), which is making more use of the Internet but

initially ran by 'snail mail'. The two current projects of most interest to 16–19 physics are *Global Warming* and *Renewable Energy*, according to the UK contact Nigel Heslop, who says that the SATIS-like learning materials have been designed to help put the 'humanity' back into science.

The *GLOBE* project is 'a worldwide network of students, teachers, and scientists working together to study and understand the global environment'.

The *European Council of International Schools* (ECIS) website has details of a venture by Warren Cookson in Oman, who 'adapted and developed three experiments which were expected to give different readings at different locations around the Earth' as the basis for collaboration between schools.

The concept of schoolchildren in different countries repeating Eratosthenes' experiment across the Internet is an appealing one, and apparently very successful too.

European Schoolnet (EUN) is looking for more schools to participate in online collaborations, one of which, between schools in UK and Greece, is looking at the effects of light pollution.

ScI-Journal is a forum for publication of students work in science. The emphasis is on developing students' ability and confidence in communicating science rather than guaranteeing scientific 'accuracy', and the coordinator Patrick Fullick does not 'censor' the articles.

A 'prize-incentivised' offer has been announced at the *APEAL* site: "Can your school create a prize-winning website which explores national and international environmental issues?"

The *STEM* and *COMO* projects at the Science Museum involve the production of web-based resources by and for educational visitors, resulting in a steadily growing database of educationally useful resources. The *STEM* Project in particular represents an attempt redefine the relationship between the Museum and its educational visitors, offering them the opportunity to express how they think its resources are best used in educational terms.

- Using the Internet can be learner-centred and highly motivating for students, some of whom are already more aware than many teachers of its likely future development

- The Internet facilitates and encourages collaboration between students in different situations and cultures

Summary

So, can the Internet support the theme of *Making Physics Connect?* It facilitates and encourages collaboration between students and other learners in different situations, cultures, across age ranges and so on. In this way a number of internationalist themes such as alternative technology and other cultural comparisons arise naturally. It encourages participants to develop their communication skills in a variety of contexts, to become more conscious of target audience, the need to communicate essential points lucidly and concisely. It necessitates development of a raft of information handling skills, most of them transferable beyond explicitly ICT contexts. Using the Internet can be learner-centred and is usually highly motivating for students, some of whom are already more aware than many teachers of its likely future development and its possible role in redefining the education process.

(An adapted version of this article is available on the Institute of Physics post-16 Initiative website, *http://post16.iop.org* , with all the references hotlinked to websites.)

DISCUSSION POINTS

- The distinction between 'information' and 'communication' is vital.

- Having the hardware to allow Internet access is a problem for some teachers, but once people must have discussed video playback in a similar way. Technology frontiers are constantly moving.

- This is a really good resource – we'll use the Initiative's Website version with its hotlinks to sites.

- The Internet is not 'good in itself'.

6 Poetry in the Physics Classroom

Peter Campbell

Only in a school curriculum are forms of human understanding divided so sharply into 'subjects'. To widen our students' horizons, we should sometimes raise the veil and reveal unsuspected connections. Here are some suggestions for exploring links between poetry and physics. However, I would not suggest you overplay this theme: you then risk upsetting your students' confidence.

A typical class might enjoy a part of a lesson spent on this theme, in several lessons scattered through a year. On the other hand, if you have sufficient interest from students, these suggestions could form the basis of an extra-curricular programme.

The language of physics

You might like to read a poem on its own, related to a physics topic you are currently teaching. Demonstrating the language of physics can inspire understandings beyond physics. I have used the following poem as part of class revision on basic electrical circuit theory.

Electricity

The night you called to tell me
that the unevenness between the days
is as simple as meeting or not meeting,
I was thinking about electricity –
how at no point on a circuit
can power diminish or accumulate,
how you also need a lack of balance
for energy to be released. Trust it.
Once, being held like that,
no edge, no end and no beginning,
I could not tell our actions apart:
if it was you who lifted my head to the light,
if it was I who said how much I wanted
to look at your face. Your beautiful face.
LAVINIA GREENLAW,
from *Night Photograph*, Faber, 1993.

Students in small groups are first asked to identify the three aspects of circuit theory described. Is the physics correct? There is one word (concept) which needs to be changed – what is it? Does the poem read as well if we make this physics 'correction'? We move on to other imagery – what does Greenlaw mean by "the unevenness between the days"? How (or why) does 'laboratory time' differ from time as felt? (A big question!)

She is also interested in the relationship between the observer and the observed phenomenon – where does this arise in physics? It might be interesting to discuss whether we think Greenlaw has studied physics. Finally, why she has chosen to use its language to convey feelings of love – new love?

Provoking discussion

A similar exploration could be done with the next poem. As revision, it would come at the end of an A level course which included particle physics and cosmology. But why not earlier in a course, to stretch the boundaries of what can be discussed in class, to reward those who have done some extracurricular reading, and to motivate others?

Pavlova's Physics

Everything in my body
has been processed
through at least one star
(except for the hydrogen)

I want to speak to you about it;
I want you to know how much
I understand - and more and more
reveals itself in waves.

I'm really a wise kid,
the kind that gets on and doesn't
need to go to college to do it,
secretly learning to peel back

the potent leaves of mathematics
while boning up on Greek at night.

For all that, the consciousness
is an outdated barn of a thing,

a slow phenomenon compared
to the speed of the senses.
Today even I'm entranced
by the submarine symmetry of my body

but, believe me, this world
is a place of bizarre consequences
where matter can appear
out of nothing and where

the light of stars is ancient
history when it gets here:
we can never understand
what we're living through at the time.

You can show me your piece of warm
thigh the length of Florida
and I'm telling you, I'm affected
by the way you look at me but I need

more dimensions than geography allows.
I'm falling forward, tumbling
into increasing disorder; yes, disorder
is increasing in the universe

and will keep increasing until
the whole shebang becomes a place
where it is remembered
only the alert rodents swam.
JO SHAPCOTT,
from *Phrase Book*, Oxford University Press, 1992.

Invoking the muse

You can also ask students to write poetry,

perhaps as an alternative homework for a special occasion. Not all of us have a flair for poetry, so encourage even terrible doggerel. You may receive something surprisingly sophisticated. Here's a student poem, collected by teacher Rupert Randall, from a year-8 pupil at Mount Grace High School.

Sound

The room was quiet without Mr Randall,
Then we heard the sound of the turning handle . .
The boy in the corner was being real bad.
And gradually Mr Randall turned mad.
He opened his mouth and began to shout
And the air vibrated as the sound came out.

The room became tense, but he made no sense
(This boy obviously must be dense.)

The air is vibrating with mutual hating.
HANNAH LEWIS
(with the last two lines shortened)

A rounded education

Once you have used poetry inspired by physics with your class a few times, I think it's good to introduce a poem which has no physics at all in it.
My justification is that the majority of my students are not studying English literature, and if I don't introduce them to new poetry, nobody will!
Choose something simple, something you like yourself, which your students will venture to understand.

Midsummer, Tobago

Broad, sun-stoned beaches.

Whiteheat.
A green river.

A bridge,
scorched yellow palms

from the summer-sleeping house
drowsing through August.

Days I have held,
days I have lost,

days that outgrow, like daughters,
my harbouring arms.
DEREK WALCOTT,
in G Benson et al., *100 Poems on the Underground*, Cassell, 1991.

Here's another.

Fog

The fog comes
on little cat feet.

It sits looking
over harbour and city
on silent haunches
and then moves on.
CARL SANDBURG,
Harvest Poems, Harcourt Brace, 1958.

If your students have enjoyed the poetry, they may be ready to compare physics with poetry. In what ways are poetry and physics similar? in what

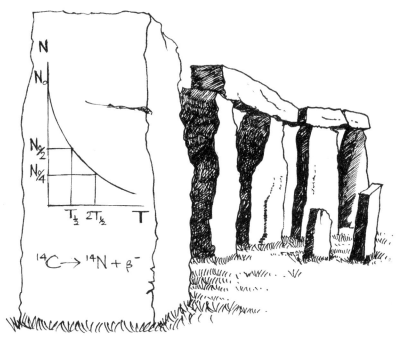

The Archaeology Lesson

ways different? These are pretty tough questions! We are getting at a big idea: are poetry and physics alike as a cultural form?

Starting points

It seems to me all of the following statements can be justified: the challenge is to find good examples, taken both from topics and processes in physics, and from the world of poetry. The class may wish to select only a few to explore.

These first three points I think are illustrated in the poems we have looked at already:

- At their frontiers, both poetry and physics aim to probe the nature of the things, to explore the unknown.
- In both poetry and physics the everyday world of sense perceptions is used to bring to view hidden realities. Patterns of images transform the objective world.

- Poetry sometimes tries to express the scientific feeling. It has also drawn on physics language to create new metaphors.

This statement might be more contentious:

- Both physics and poetry sometimes grasp to understand our place in the universe.

And this one may need the wider knowledge of an English specialist to decide – so why not cooperate with an English teacher to stage a discussion before a student audience?

- Both exploit the unconscious mind in creating beauty, playing on the boundaries between symmetry and asymmetry, between patterns and pattern-breaking. Both have their eureka moments, when new links are made between ideas.

In discussing the following, students are likely to learn something new about the nature of science:

- In their use of language, both physics and poetry aim for precision, but physics avoids ambiguity where poetry exploits it to give the reader scope for interpreting meanings.
- Science is often done collectively, and at least in public papers expresses its findings impersonally. By contrast, poetry is done alone and expresses itself personally.
- Both are practised internationally and can defy authority.

Souces of inspiration

Finally, here are some more sources of science-related poetry. I wish you and your students much pleasure in discussions of this sort.

John Carey (ed), *The Faber Book of Science*, 1995.

Bernard Dixon (ed), *From Creation to Chaos*, Blackwell, 1989.

Timothy Ferris (ed), *The World Treasury of Physics, Astronomy and Mathematics*, Little, Brown, 1991.

Walter Gratzer (ed), *The Longman Literary Companion to Science*, 1989.

Derek Jones (ed), *Rhyme and Reason*, Channel 4 Television, 1995.

DISCUSSION POINTS

- Linking with article 3 (on Faraday), poetry can draw attention to the use of language in science. Both science and vernacular terminology move on, so language as used by the scientist no longer has the same resonance with common usage; for example, 'charge' and 'capacitance' were clearly linked with the idea of electricity as a fluid.

- Sometimes students without English as a first language are at an advantage because they don't appreciate the additional meanings of specific words used in physics.

- Scientists themelves some times speak poetically; for example, Dalton compared numbers of atoms to the number of 'stars in the firmament'.

- Physics has also affected structures of creative writing this century. Durrell's *Alexandria Quartet* has three books located in space, while the fourth provides the time dimension.

- Physics educators have a responsibility to promote a more attractive image of physics.

7 Famous Physicists

Peter Campbell

DISCUSSION POINTS

- Are all physicists dead, white males?

- Isn't the notion of 'famous physicists' a trap we should avoid? What about ordinary, unfamous physicists? What about engineers? What about the importance of team work in most science today?

Thales	James Watt	Maria Goeppert-Mayer
Democritus	J P Joule	Ernest Rutherford
Aristotle	Georg Simon Ohm	Rosalind Franklin
Archimedes	James C Maxwell	Abdus Salam
Lucretius	Heinrich Hertz	I I Rabi
William Gilbert	Henri Becquerel	Paul Dirac
Galileo Galilei	J J Thomson	Hideki Yukawa
Johann Kepler	Carlo Rubbia	Gerard 't Hooft
Isaac Newton	Albert Einstein	Emmy Noether
H C Oersted	John Wheeler	Chen Shiung Wu
Charles Coulomb	Marie Curie	John Bell
Robert Boyle	Lise Meitner	Werner Heisenberg
Robert Hooke	Richard Feynman	Jocelyn Bell-Burnell
Thomas Young	Enrico Fermi	James Chadwick
Sadi Carnot	Stephen Hawking	Chandrasekhara Raman
John Dalton	Tsung Dao Lee	Erwin Schrodinger
Michael Faraday	Chen Ning Yang	Peter Kapitza
Laura Bassi	Hans Bethe	Bernard Lovell
Alessandro Volta	Edwin Hubble	Henrietta Leavitt
Caroline Herschel	Andrei Sakharov	Alain Aspect

Students have probably learned a few names in physics, if only because SI units are named after scientists.

But do they know any early names, indicating the deep roots of natural philosophy?

We are still asking the ancient questions:

- Are there patterns in the world?
- Can we explain, in simple terms, the many forms of matter and changes of form?
- Can we know more about our context – our history, our future and the Universe which surrounds us?

Do students know of more than one or two of the great twentieth century physicists?
Do they realize how a largely European science has been internationalized?
How women have contributed, especially in astronomy?
Do they know there are more physicists living today than physicists from all previous times?

Ask students to choose a scientist and research his or her life and work, for presentation either orally or in writing.

Written accounts can be collated into attractive display posters for the laboratory wall.

8 Women in Physics

Lyn Branson

I love the world I live in and I love to explore and try to understand the way it works. So physics was a natural choice for me.

I was lucky, I think, because I went to a single-sex school where I was not concerned about my image with my peers – we were all female anyway. My mother, thankfully, had no problems with my choices of maths and physics, and my father was equally encouraging. It could have been very different.

Picture a young woman, just getting in touch with her sense of her femininity, her sexuality. Slowly the young males around her take on a new significance, and both sexes start to explore their new identities and relationship to one another. There is perhaps a strong need to become stereotypical of one's sex, or at least to try to discover what *is* stereotypical.

Making important choices

Soon, or worse, at the same time, this young person will need to make important life-defining choices about her future. Or at least this must be how subject choices at 14 and 16 must feel. She looks around at the females in the media, her teachers, her parents and their friends.

Her physics teacher is probably male and often unknowingly biased in his treatment of the young women in his classes. Concerned teachers may undermine young female physicists by a too solicitous approach which can be as damaging to their self-confidence as direct discrimination. Languages, humanities subjects, the arts are all populated with overtly feminine women, discussing and working together, an image which appeals to young women.

The biology teacher may be female, this will seem more appealing for females with a scientific interest. Also biology seems to lead to jobs and careers where there is an obvious connection with people; a chance to move into an adult world comfortable with this new femininity.

Not so physics. The whole subject area is massively populated with males, and men have a different way of being with each other and their lives than we women.

Around twenty years ago, when I had my qualifications in physics and had been teaching for a few years, I was pretty smug about sexual equality. I felt as if I could do any job I really wanted to, and thought that we women had levelled the pitch. An article in a Sunday newspaper set me thinking.

The article pointed out that we lived in a society that had been shaped by men, and that the women who were succeeding were in the mould of this society. This was not real equality, but women shaping themselves to fit into a man-made structure, and in doing so either neglecting part of themselves, or fitting more than one role into one life.

Either way it was not healthy. Suddenly I saw, with sickening clarity, how far our society needed to go to evolve into one of true equality for both sexes. It was a very significant moment for me and I sometimes feel that we haven't moved very far down the road. We are moving however.

But we have a particular problem with physics. The subject of physics, perhaps largely because it has been developed by men, has a very masculine feel to it in every way.

Confronting the stereotypes

Many people asked to describe a physicist will paint a picture of someone a bit out of touch with day-to-day life and people; a man, probably old, and beavering away in isolation. This is not a true picture, but it is out there and needs changing if young women are to feel that the subject is as much theirs as their male peers'.

The stereotypical image of physics draws many young men naturally towards it. I am, of course, talking in absolutes. I am well aware that there are many men in the world of physics who would not fit any of these stereotypical images, but the overall image in the world out there, I fear, is stereotypical, and uninformed.

After a few years teaching I was becoming disenchanted

- Physics has grown through the thoughts and observations mostly of men

- If we could achieve an equal presence of women I am sure that our perceptions of the physical world would shift

with my subject and nearly gave it up. I felt that I needed to be able to bring more of myself into what I was doing, and somehow this seemed impossible with physics. I found myself feeling uncomfortable, constrained. I was happier in tutorial work where I could relate more closely with the students.

I explored counselling and creative photography, obtained qualifications in both and incorporated them into my role at college. I gained the strength to start to change my attitude to my teaching. This involved becoming myself in the teaching, rather than bringing parts of myself to the teaching – a very scary but rewarding process. I also began more and more to give my students ownership of their learning. I am now very informal: sharing the learning experience with my students; working on my relationship with them and their relationship with themselves as well as teaching them about physics.

Sharing perceptions

So how do we bring about a change in the gender balance in physics? I think slowly and with great effort, but this effort is worthwhile. While there are many young women out there who potentially could enjoy this subject, physics needs them and not just as numbers. Physics has grown through the thoughts and observations of mostly men – this needs broad-

ening. I have no idea what will change, or how, but I do know that the two genders think and work very differently. But if we could achieve an equal presence of women, then I am sure that our perceptions of the physical world would jointly shift.

We need to support and encourage the young women who are interested. If there are not enough qualified female physics teachers out there, we need to reach out to the students in any way we can to help them see that physics welcomes them and offers a place to them.

A way forward

The Internet seems an obvious mechanism for doing this. As the use of this communication superhighway increases, we can initiate websites which will allow young female physicists to ask questions and gain information in a safe way and take the commitment as far as they wish. A web of women physicists which could encourage and share.

This is the way women work best, I think. We need to bring more humanity into our teaching of the subject, more co-operative ways of learning, more group identity, more support of every sort.

Young girls expect to find physics difficult; this fits with their perception of femininity. They need to see feminine role models in physics like Linda Williams, a cabaret artist who sings and dances about physics! Sadly she is American, as are most of the web sites I have found around women in physics – some links are on the Institute of Physics website.

We need to become far more active in encouraging, indeed expecting, females to be part of physics. Who knows where our understanding of our Universe may go if the genders share its exploration?

Further reading

Some readers might be interested to read Sadie Plant, *Zeros and Ones* Fourth Estate, 1997. Described as a 'manifesto on the relationship between women and machines', it threads an account of Ada Lovelace with views about women and networks, computing and the Internet.

DISCUSSION POINTS

- Young women are more likely to ask questions in class – but they perceive this as a weakness, and it adds to their sense of insecurity.

- Will this article make men reflect on their own behaviour?

- Some women physicists deliberately flout the male culture of the workplace.

- The physical environment of the physics lab is usually unattractive, and puts young women off.

9 Presenting Physics

Lawrence Herklots

New entrants to the teaching profession are offered many pieces of worldly wisdom from their more experienced, if not to say jaundiced, colleagues.

The two pearls repeated more than most are 'don't smile until half term' and 'you don't really know you subject until you teach it'. The first piece of advice is taken to heart by many who continue to favour a facial expression that hovers between a grimace and a sneer far beyond the half-term deadline. The second statement is, reassuringly, often found to be no less than stating the truth.

How do we learn?
Teachers of physics are individuals and learn their physics in individual ways. Many, perhaps the majority, reach understanding through mathematical models, whereas others find that experimental work locks the ideas into place.

When faced with a class of twenty or so other individuals, the teacher's understanding of the subject is probed because the questions that are raised by students come from many different perspectives. The teacher comes to 'really know the subject' by combining these different perspectives. This process is the making of many teachers, but the initial stages can be more than a little harrowing.

When students are given the opportunity to present aspects of physics to their peers, they are placed in a similar situation to the fledgling teacher. A student can gain a huge boost to his or her confidence by successfully describing, say, Geiger and Marsden's work on alpha-particle scattering to the class and then answering questions with knowledge and understanding. But not all sixth formers relish this opportunity and it is simply bad teaching to put students off physics by forcing them into what may be fairly (though inelegantly) described as 'inappropriate ancillary activity'.

The trick is to give students the opportunity to present physics from their own perspective without, necessarily, subjecting them to an interrogation from the class. This can be achieved by not only encouraging all students to make a presentation of some form or other, but also by broadening the scope to include lectures, poster presentations and web pages.

Student lectures
A level students have been giving lectures in classes for many years. Having been encouraged to investigate a particular area of interest linked to the physics studied, they present a short talk to the rest of the class. Overhead projector transparencies can add illustrations and illustrative calculations. Typically, a talk will last around ten minutes with another ten minutes of questions.

The teacher needs to guard against these talks developing into rather desultory affairs in which the students read out second-hand physics that is not really understood. This benefits nobody. However, giving students a little direction at the same time as giving them the freedom and confidence to investigate areas of real interest, can make the finished talks refreshingly individualistic and of real value.

A good student lecture can be the highlight of many weeks' work – students often ask more pertinent and more wide-ranging questions of the 'king for a day' than they do of the teacher. The reactions of class teachers range from admiration through to surprise and enjoyment at the range of talks produced by the students.

Here are just a few examples
The study of materials provides many possibilities for student lectures. One memorable example was a discussion of composite materials which involved the destruction of ready-prepared jellies containing different fibres and lentils as composites.

There is nothing new in this, but the presentation was wonderful. After a fairly pedantic opening description, out came the jellies and the mallet. The strengthening effects of

> - **The illustrations would have made Heath-Robinson proud, but the application of physics showed great understanding and imagination**
>
> - **Designing a poster involves sifting many pages of text to discover the essence of an idea**

aggregates were shown with great clarity.

Having been introduced to basic wave theory, a recent student of mine produced detailed plans for the ultimate luxury bath in which waves of warm water caressed the bather with a frequency digitally selected with a control pad. What fascinated the student and his audience was the challenge of mathematically modelling these oscillations. Luckily they approximated to simple harmonic motion.

I remember that the driving force for the waves was a large oscillating titanium duck. The illustrations would have made Heath-Robinson proud, but the application of physics was serious and showed understanding and imaginative application.

As an example of a more formal talk, the student who chose 'The Early Years of Radioactivity' gave the class much more than a chronology of the development of simple atomic theory because her teacher had asked her the simple question, 'Which came first, theory or discovery?'.

This gave her a position to defend – she believed that experimental discovery led and stimulated theoretical advance. This, in turn, led to a vigorous discussion amongthe class which brought many preconceptions into sharp focus.

Poster demonstrations

Science departments often display posters produced by pupils at Key Stage 3 or 4. A level students have fewer opportunities for this type of presentation. This is unfortunate because it limits creative opportunities and, subliminally, it suggests to younger pupils that A level physics is a rather dry affair.

A poster demonstration is a development of the student lecture. Instead of using an overhead projector or experimental demonstration, the students talk their way around the poster they have prepared. This gives free rein to students' frequent interest in graphic design but still needs a clear understanding of the physics concerned.

This approach can be particularly useful in dealing with historical aspects of physics. Posters using some form of timeline encourage the eye to take a 'straight line' route when reading of linear, accretionary development. But when ideas

start pouring in (for example, in the aftermath of a Kuhnian paradigm-shift), the timeline explodes like a fireworks rocket. The importance of key ideas and experimental results can be shown with great clarity.

The posters take at least as much preparation as the short lectures. The process of designing the poster involves sifting many pages of text to discover the essence of the ideas being considered. Students are less inclined to waffle on a poster, and the individual nature of poster design discourages the word-for-word copying that bedevils some student lectures.

This is seen as fun and rather less intense than the usual diet of theory and practical, but it is not a soft option. When the teacher and students discuss the poster during preparation, there is an excellent opportunity to probe the student's understanding of the basic physics. And of course, when the demonstration is over the poster can be displayed to all and sundry, giving a gentle hint to those who walk by that, perhaps, that A level physics is not as esoteric a discipline as they may have thought.

Web page design

The relative ease with which web pages can be constructed offers rich opportunities for channeling students' interest in the possibilities of using the new technology and producing a very useful student resource.

At its most basic, a web page is little more than a poster on a computer monitor with a few little spinning things in the margins. Although this looks attractive and students take great pride in constructing a page, much more can be done. There is little real use in a page that simply gives a few facts about, say, the electron shown in a particularly garish purple against a wallpaper of repeating images of JJ Thomson. It is the interactive nature of web pages that needs to be utilized. The students that began with a wallpaper of Thomson clones soon developed the page into something altogether more interesting.

Their final product, 'The Electron: an Intimate History' leads the reader deeper into the developing understanding of the electron whenever the question 'Do you want to go

- It is the design possibilities of web pages rather than access to a potential audience of billions that excites students

- The relative ease with which web pages can be constructed offers rich opportunities for channeling students' interest in using new technology

further?' is answered 'yes'. These students (at the time of writing in their second year of A Level) are further developing the interactive nature of the pages by writing questions which need to be answered correctly to allow the reader to progress. The questions are being written by all members of the teaching set and range from simple to difficult, trivial to fundamental.

When completed, this work will be a resource the whole department can use. A bank of pages is in the early stages of production and will be used in all areas of the syllabus and provide a useful independent learning resource.

Designing web pages is relatively simple. Students enjoy the creative nature of the work but also become concerned that the physics is clear and correct. This provides more opportunities for the teacher to work on the fundamental principles of the subject with the students.

The finished pages need not go on the Web. Many schools and colleges have networks that can support a form of 'intranet' which is ideal for this kind of work. It is the design possibilities of web pages rather than the access to a potential audience of billions that should excite the students. They may, however, take a little persuading!

Conclusion

Physics is creative, difficult, precise, imaginative, personal, objective and almost everything else. We all know this. We also know that these characteristics arise from the fundamental ideas and methodology of the subject.

But students won't know the richness of the subject until they are given the chance to use the 'fundamental ideas and methodology' in areas that interest them and in ways that excite them. It is not enough for the small minority who progress to university to glimpse the richness of their chosen subject. Physics is big enough, surely, for all 16–18 years olds who study it to find a perspective that brings the subject to life. One way of encouraging engagement with the wider sweep of physics is to give students opportunities to present ideas to their peers. Perhaps this will become a feature of some assessment schemes, perhaps not.

Assessed or not, student presentations in all their manifestations can be beneficial to the student, to the class and to the department. They can be fun, as well.

Further Information
The web pages written by students this year can be accessed at:
http://www.kes.hants.sch.uk/
Hints about contructing web pages can be found at:
http://www.kes.hants.sch.uk/webdesign/

DISCUSSION POINTS

- Presentations take time, but they are a valuable learning tool and should not be seen as an added extra.

- Teachers need INSET which includes advising students on presentation skills, and how to run oral presentations while interfering minimally.

- Partly because they give choice, presentations encourage student engagement with their topic and thinking about synthesis.

- Presentations of all types provide a useful assessment tool for differentiation.

10 Physics in Advanced GNVQ Science

Lyn Branson

In GNVQ Science we can choose the content of our own assignments to a certain extent, so long as we cover all the theory required. This enables us to build on areas of interest and expertise within the student body and the locality.

For those who have not yet become involved, I will begin by explaining briefly what a GNVQ Advanced course entails.

There is a framework of science theory, which is tested in short multiple choice exams; however, the bulk of the work, and the basis for the grade awarded, rests with numerous coursework assignments which students put together as a portfolio.

The topics which need to be covered for successful completion of the course are defined, and there are test specifications to clarify what is to be included in the mandatory tests. Students study two mandatory Units in each of the three main disciplines of biology, chemistry and physics together with two more general Units. The twelve Unit Advanced GNVQ (equivalent to two GCE A levels) is then completed by adding four optional Units from any of the three sciences. The students are also assessed on the key skills of communication, IT and numeracy.

Making physics relevant to everyday life

Having taught the physics Units for about five years, I was asked to write about Advanced GNVQ Science. I found myself unable to begin without consulting my students. I think that starts to say quite a lot about what I find different and very valuable about this course compared to GCE A levels, which I have taught for twenty five years.

While I have changed the way I deliver the GCE A level physics content – now aiming to facilitate the students' learning of the syllabus – I feel that it is still possible to teach GCE A level from a pretty didactic and formal base. Not so Advanced GNVQ.

For example, collision processes would be studied in both GCE A level and Advanced GNVQ. In the former, student activity would likely be limited to theory and calculation. In Advanced GNVQ the students might be set an assignment to research safety mechanisms in different modes of transport relating to rate of change of momentum, and give sample calculations using their found data.

This assignment, assessed as part of one of the mandatory Physics Units, gives the topic far more relevance and life. An essential part of the course, this learning process encourages far more joint endeavours between teacher and the student group.

Ask the students

So I felt that I had to go and ask my students what they felt was the difference, and what was good about Advanced GNVQ compared to GCE A level. All of them have friends studying A level and are well aware of their friends' experience. I found myself little surprised by, and entirely in agreement with, what the group said.

They find that they gain confidence in themselves quite quickly on the course because they have to find out about things for themselves, and share information among the group very early on. No possibility here of being a loner! They very soon hone their IT skills so that the work presented appears looking very professional. Indeed many of them can teach me quite a lot about using computers effectively.

They are very proactive, and soon learn to access a variety of sources of information. Often they will choose to contact industry for information as part of an assignment, and so gain confidence in interacting with the world of work. They are credited for their IT and communication skills, something which promises to be included in the new GCE A level schemes.

One of the major benefits of GNVQ is that it is a whole learning programme. The group is together in every class

INSTITUTE OF PHYSICS

- **Students often choose to contact industry for information as part of an assignment, gaining confidence by interacting with the world of work**

- **One of the frustrations is that the strong GCSE candidates usually still turn to GCE A levels**

throughout the week. Again, very different from the GCE A level student who is unlikely to come across two groups containing identical students. If they do, many GCE A level teachers – and perhaps most notably in the sciences – will allow them to select the individuals with whom they interact. They can be quite isolated if they choose.

This is not possible in Advanced GNVQ, because working as part of a group is integral to the course. I was gratified that the students valued this as much as I do. We often have some difficult times in tutorials during the latter part of the first term, when they each start to gain confidence in themselves and lose patience with each other.

A solid preparation for the world of work

With help and support they work through this each year, and become very businesslike in their approach. This must be invaluable preparation for the world of work where, as we all know, we have to learn to work effectively alongside people whether we would choose to or not. I think that this is one of the reasons that they are so eminently employable when they finish.

Students are, however, equally able to go on to university if they achieve the right academic standard. They are used to finding information out for themselves, and have learned to plan their time effectively.

This is one of the qualities we assess as part of their portfolio of assignments. I don't see the same level of skill developing in the GCE A level students across the board. We also assess GNVQ students' ability to evaluate what they have done, to reflect critically on the way they have approached an assignment, and make suggestions as to how they could tackle the thing more effectively if they were to do it again.

Of course not all Advanced GNVQ students are high flyers. Indeed one of the frustrations is that the strong GCSE candidates in the main still turn to GCE A levels. I hope that will change, but it is a bit like a chicken and egg situation. We need some really strong candidates to come and exploit the GNVQ approach to its full extent. The scope for them to really take off is limitless. A candidate achieving Advanced GNVQ graded at Distinction level would be the equivalent of an A or B at GCE A level.

Fortunately higher education is beginning to understand and accept Advanced GNVQ. All of our students who wanted university places have been successful over the time we have been running the course.

I feel quite strongly that there is a need for this type of course as part of every students' post-16 education. We are moving towards a five-subject first year of post-sixteen study. I hope that in future sixteen year olds will be compelled to combine some vocational (GNVQ) experience with AS levels in the first year. Then perhaps we can give *all* our youngsters the benefits of this type of educational experience.

DISCUSSION POINTS

- Because students are together as a teaching group all the time, the GNVQ teacher must ensure positive group dynamics develop.

- A level students are sometimes envious of the practical project work they see left behind in the lab by GNVQ students – things like building and testing solid fuel rockets.

- GNVQ students learn much more about what scientists do than the standard A level student will.

- Presentation skills develop naturally in GNVQ Science. The fact that presentations work successfully in GNVQ shows they benefit not only the most able students.

11 Connections in the NEAB Turning Points Option

Philip Britton
& Sarah Cramoysan

For the last few years the NEAB has offered module PH09 'Turning Points in Physics' as one of the nine modules which make up the A level syllabus.

Students (or their teachers) choose to study three of six optional modules of which this is one possibility. The module can also be offered as one of the four possible option topics for 'end-of-course' exam students.

In the preamble to the module we read: "[it aims to] enable key developments in physics to be studied in depth so that students can approach, from an historical viewpoint, the significance of major conceptual shifts in the subject, both in terms of the understanding of the subject and in terms of its experimental basis. Many present day technological industries are the consequences of such key developments and the topics illustrate how unforeseen technologies develop from new discoveries."

Since, in both our schools, students take the end-of-course exam, teaching this option comes towards the end of a two-year A level course when students have fairly well developed skills in physics but are at the stage when, despite the increased motivation of the approaching exams, enthusiasm for physics as a subject rather than an examination can be waning.

Putting physics in a living context

It has provided an ideal opportunity to set physics once again in a human context, to discuss the people who have made science, to assess the way scientific ideas are shaped by human nature, and how human lives are shaped by those scientific discoveries.

The option has four sections: 'the discovery of the electron', 'wave particle duality', 'special relativity' and 'towards absolute zero'.

The first two provide a particularly good opportunity to discuss the figures who 'made' this science. One successful teaching idea has been to copy pictures of apparently diverse scientific figures onto a handout, each name known to the class for other work. A lesson is spent giving a class a thumbnail sketch of each person in human terms and spinning a thread of the story of scientific adventure that links the people together.

Kelvin's involvement in debate with Huxley and Darwin, his insistence on applying his thermodynamics to the dating of the earth, and dare-devil exploits with the trans-Atlantic cable provide a suddenly three-dimensional view of a rather odd figure. Otherwise students may wonder how a temperature scale has come to be named after a suburb of Glasgow. This tendency for scientific figures at the turn of the century to be polymaths provides a thought-provoking opportunity to discuss the structure of the scientific community today.

Physicists are human too

As discussion develops, students are at first amazed that scientific ideas and advances, which they have tended to view objectively as either right or wrong, can be held back through lack of acceptance due to vested interests and personal differences.

Is it a correct reading of history to presume Thomas Young required French backing for his wave model of light to convince an England still in awe of Newton? Did Kelvin really tell Thomson that his new recruit Rutherford should forget about his ridiculous early research on radio that would come to nothing?

It is perhaps only natural that the recurring theme of an old guard failing to accept new ideas strikes a resonance among sixth-form students. This theme is particularly easily developed when discussing wave particle duality.

Equally the links between well known figures – where they worked, who worked with whom, the suspicion of friendly rivalry and the suspected lively banter and idle chatter – has lightened many of these summer term A level classes.

We study the JJ Thomson research group (perhaps better known as the moustache photograph!) and wonder: who if anyone cracked a joke as the camera flashed; who had been

- **At first students are amazed that scientific ideas and advances can be held back by vested interests and personal differences**

- **It is perhaps only natural that the recurring theme of an old guard failing to accept new ideas strikes a resonance among sixth-form students**

late to arrive; who went off afterwards to have a drink with whom? For us it is these five-minute interludes of idle speculation that make the option an enjoyable experience, Hopefully they are not entirely trivial in the sense that students are being encouraged to think of physicists as human, even if that is male and moustached humans.

Such wild speculation, grounded in accurate if limited information, brings the topic alive. The move from the electron as a particle to the electron as a wave can be lightened by pausing to imagine the father and son conversation between GP and JJ Thomson. Did JJ, involved in the discovery of the corpuscle that became known as the electron, have sharp words for GP for his part in showing electrons diffract, a clear indication of a wave-like nature?

The idea of research groups and the development of that concept from the small gathering of Thomson, through Rutherford's much larger Cavendish group, to the vast group photographs from CERN allows comment on the changing nature of scientific endeavour.

Anecdotal evidence

We have made reference to limited historical detail, and of course anecdotes become apocryphal rather than accurate, so there is a constant need to check references and a great need to share and collect stories and resources.

It is perhaps important to realise that we are not teaching the history of physics, but setting the examined content of some basic calculation work on electron physics in an historical and human context.

But where do anecdotes come from? For us this is the crucial question, the answer to which will either support or under-

mine this teaching approach. Obviously they come from personal reading, both authoritative scientific histories and more popular accounts. They come from magazine articles such as occasionally appear in *Physics World* (for example: on Dirac in February 1998, or on Rutherford in September 1998). They come from hearing lectures, coffee time chats with other physicists and conference evenings in the bar.

But few have time for all of that. It can be rewarding and fun for the teacher to research, but there are times when an off the peg anecdote would be nice. Or given the time to read one article, having a good one recommended would be useful. There is a need to share ideas and support each other here. We hope the Institute of Physics post-16 Initiative will encourage this, through both local teacher networks and the Institute's website.

Although there is the opportunity to mention the development of scientific ideas when teaching the discovery of the electron and wave particle duality, the section on special relativity is especially good for this. Clarifying the line between valuable background and digression and the examined content is vital if students are to be enthused and not feel overloaded.

While teaching, it is important to identify that, in terms of examination questions, the content is fairly straightforward and limited. Given the context, the increasing worry about non-relativistic physics, the various suggested solutions which amount to special relativity but lack its conceptual advance, the implications of having no absolute reference frame are all areas for developing discussion. Not to mention the persistence of Michelson and Morley and the supreme importance of being sure that nothing happened.

'What did the dog do in the night time?'

This drama must prod the conscience of any sixth former who has complained of nothing happening in their own experiment (even if in school it probably is because they haven't plugged it in). Who can fail to delight in the anecdote about the world's most famous sleuth Sherlock Holmes who,

JJ Thomson's research group photographed in 1897

- **Here we are not teaching the history of physics, but setting the examined content in an historical and human context**

- **Teaching this option can be a delight and the enthusiasm of otherwise jaded and world-weary A level students is its own reward**

when investigating an evening abduction, remarked: "What did the dog do in the night time, Watson?" "The dog did nothing in the night time." "Precisely!"

Moving through to the final section of the option we have the opportunity to bring the human endeavours of physics up to date with superconductivity and superfluidity.

Not only are the people who discovered the electron real, but so are those working in physics today. They went to schools (which often still exist), they studied for A level, they have families; many have children of sixth-form age.

The reliance of these scientific advances on previous ideas and the technological development of vacuum and cooling technology can be drawn out. So may the enormous implications, both already existing and imagined, of high-temperature superconductors, the rapid pace of development and the huge financial sums involved. Yet, on the other hand, the abstruse and weird and wonderful theoretical physics being developed to try to explain, and so understand and develop our use of, these phenomena can be seen.

Then there is another chance to empathise. Which sixth-form student faced with Ohm's results of zero resistance would not have quietly returned the broken meter to the apparatus trolley?

Teaching this option can be a delight. Drawing out the historical and human connections cannot take too much time, perhaps ten minutes in each 50 might be given over to these context setting discussions and interludes, but the enthusiasm of otherwise jaded and world weary A level students is its own reward.

And what of future developments? For us teaching this option brings its own duality. The context of teaching is necessarily distinguished from the examined content. The option is examined through about nine structured questions. These are often calculations (the wavelength corresponding to a certain energy) or recall of fact (the postulates of special relativity).

There are some questions which require description of historical results and even some, although few, requiring limited analysis – for example the move from light as particles to waves). Perhaps future developments in examining will allow the more speculative aspects of the module to be examined. Or will the act of observation destroy the nature of the teaching style being used?

DISCUSSION POINTS

- It is very important to have, as we have here, real examples from the classroom told with wit and insight. They can help teachers to see how to manage something similar for themselves.

- The photograph has more to show than moustaches. Don't their clothes look dreadful? This owes something to fashion, but probably more to the lack of synthetic fibres. Old photographs tell hidden tales of technological change: a striking example being the famous photograph of Isambard Kingdom Brunel posed in front of the *Great Eastern*'s anchor chain.

12　Nuffield Advanced Physics 'Research and Analysis'

Lawrence Herklots

... an opportunity for good practice

Introduction

The 'Research and Analysis' (R & A) paper was first used as an assessment in the 1992 examination. It requires candidates to write a paper of 2000 to 4000 words on a subject based on ideas from one of the option units of the Nuffield A Level course. The option units are:

- digital electronics
- energy sources
- linear electronics, feedback and control
- energy and entropy

This is a very broad range and allows candidates to look into almost every area of physics.

The papers produced by the candidates are assessed on five criteria:

- interest and independence
- range, variety and relevance of sources used
- amount, accuracy and relevance of physics used
- analysis
- communication

Good projects are interestingly written, show good understanding of physics and have the stamp of the candidates' own personality and interests.

How the 'Research and Analysis' paper encourages good practice

There are two strands that form the basis of the R & A task. The first strand is the application of rigorous physics to novel situations that are chosen by the candidates themselves. The second strand is the writing of the document in a clear, interesting and attractive manner.

The R & A paper is a real opportunity for candidates of all abilities to produce a meaningful and interesting document rooted in carefully assessed physics. This produces a differentiating assessment task which further encourages candidates

in their efforts and, crucially, makes the input of the teacher invaluable.

The enthusiasm shown by the candidates in their written reports is more often than not a reflection of the seriousness and interest with which teachers discuss the candidates' ideas. The R & A time is not a fallow period for the teacher but an opportunity to teach on an individual basis that the pressure of the syllabus rarely affords. As all physics relies on fundamental principles, it is often found that candidates develop understanding of some of the foundation ideas of the syllabus more clearly through working on R & A.

There are three phases to the candidates' experience of R & A. These are:

- research and literature review
- analysis
- writing

In each of these areas the candidates consider physics in novel situations. As the project progresses the candidates' sense of ownership of the project increases and individual interests, flair and character show through. To understand how this can develop it is worth considering the three phases in more detail.

Research and Literature Review

Having agreed an area of study with the teacher – which can be anything from black holes to engine management and back via NMR scanning and the history of the transistor – the candidate reviews the literature and makes use of local contacts. It is not uncommon for candidates to visit universities and hospitals to interview those actively engaged in research.

Often projects have a local flavour. Candidates in Southampton, for example, often produce projects concerned with geothermal energy, whereas Cornish candidates more frequently consider wind energy. When the research phase includes personal contact, the candidates get a glimpse of the application of physics in industry and research.

- Good projects are interestingly written, show good understanding of physics, and have the stamp of the candidate's own personality and interests

- R & A offers an opportunity to teach on an individual basis which the pressure of the syllabus rarely affords

This is fertile ground in which to plant the idea of a career involving physics. To adapt an old proverb, 'one personal contact is worth a thousand glossy recruitment posters'.

Of course, some projects do not lend themselves to personal contacts. For example, cosmology is popular among candidates, but relatively few cosmologists are around to be interviewed. Thus the research experience for these candidates is rather different. It involves scanning texts, focusing and honing ideas into more defined areas and looking into those areas in greater detail and depth.

There is a clear role for the teacher in developing these important skills. In the process candidates discover the pros and cons of different text-based information sources and get the opportunity to read some first-rate technical and popular science writing.

Analysis

In the second phase the candidates face the task of analysis. In R & A this involves teasing out the physics of the subject and using ideas met in the Nuffield course to explain the phenomena, technology or process that they are describing.

There is, once again, an opportunity for individual teaching here to ensure that the candidates have as good a grasp of the physics as practicable. Well-tutored candidates often include example calculations in their reports to show the application of physics to the topic. This can be as straightforward as considering the temperature rise in the water in a geothermal station to the more esoteric and mathematically advanced modelling of deep ocean currents from a thermodynamic perspective. In R & A, as in all else, 'you cut your cloth to match your means'.

Writing

As the project proceeds the focus shifts to communicating the ideas in a clear and concise manner. A good prose style is not enough. The writing phase also involves careful, embedded referencing and employing a variety of diagrams, charts and text.

It is at this stage that the individuality of the candidates begins to show. Good candidates aim to catch the interest of the reader rather than presenting a list of facts and theories. In such candidates' hands the project becomes a piece of work that the assessor reads for pleasure instead of duty. Indeed, the quality of the best candidates' work can be quite astonishing, certainly of high enough quality to use with future candidates as both exemplars and background reading articles.

The R & A project encourages the application of physics to a novel context. The range of possible titles is limited only by the imagination of the candidates and their teachers. Of course, hopeless titles are not accepted by the teachers and in such cases the candidate would have to think again. Often, however, candidates suggest ideas which really are surprisingly novel and yet workable.

Since the candidates can choose their areas of study, a natural filtering occurs: those with practical interests choose technical/technological aspects of physics and candidates whose interests are more theoretical choose areas like cosmology, astrophysics and quantum physics.

The R & A project doesn't attempt to constrain individual interests or mould all young people into theoreticians or technologists. Its aim is to encourage candidates to explore the physics that interests them. As an examiner who each year reads a wide variety of scripts, it is clear to me that R&A projects often succeeds in this respect.

Teachers are straining under the load of marking and paperwork that coursework components in 11–18 education have produced. R & A is coursework and needs careful assessing. However, it is not overly burdensome because most of the assessing is carried out during the lesson time given over to the component. The more the teacher talks to the candidates about their ideas the easier it is to assess the finished articles .

This cannot be overemphasised. By discussing projects with the candidates the teacher gains a clear understanding of how things are progressing, shows a real and pointed

- **In some candidates' hands the project becomes a piece of work that the assessor reads for pleasure rather than out of duty**

interest in what the candidates are doing, helping them to focus their ideas. Everyone wins. It is not surprising that many teachers each year look forward to the R & A component of the Nuffield course.

With the development of similar components in the new Advancing Physics AS/A2 course, my hope is that many more students will taste the excitement of finding out for themselves, and many more teachers will be amazed at how much 16–18 olds can achieve.

DISCUSSION POINTS

- Research techniques and writing skills need to be taught in class.

- The teacher is not in the usual role of being in charge and knowing everything.

- Well-managed, this activity allows teachers to give one-to-one support to students.

13 Salters Horners Advanced Physics

Elizabeth Swinbank

Like all science courses with the Salters brand name, Salters Horners Advanced Physics is context led.

The structure and content of the course are driven by contexts, applications and issues which lead students into the exploration of scientific content and the development of skills. This contrasts with the more conventional approach in which real-life examples are used as incidental embellishments or appear only after a prolonged exposition of content – if, indeed, they feature at all.

The context-led approach is seen as a way of building in 'connections' as an integral part of students' learning, providing motivation and engaging interest by demonstrating the up-to-date nature of the content and its relevance to students' own experiences, interests and aspirations. It has been used successfully in the Salters Science and Advanced Chemistry courses which have been running for several years.

The Salters Horners Advanced Physics course is at an earlier stage. The course began as a pilot in September 1998 with a cohort of a thousand students (the maximum allowed under QCA rules). Feedback collected during the pilot will help point the way forward for future developments involving Salters and other courses.

Physics in context – making connections

The Salters Horners course makes use of contexts in various specific ways, all of which have some relevance to the general aim of making physics 'connect' both to the world beyond the school laboratory and internally as a coherent discipline. The effectiveness of all of these is being evaluated as part of the pilot.

People and physics

The context-led approach enables people to feature promi-

nently in the course. For example, the course materials include an interview with an environmental health officer, an article by a young space engineer describing his work, and an account of a theoretical physicist hitting on the idea of cosmological inflation. Such devices are intended to demonstrate the importance of human endeavour in scientific development and illustrate some of the many careers and areas of further study that involve physics.

Context-related activities

Setting the course materials in context leads readily to the inclusion of student activities that connect to present-day issues and applications. Some of these take place on the laboratory bench – for example, learning about and using polarimetry and refractometry in order to determine the sugar content of syrups for use in the food industry – while others involve ICT, as when students use the Internet to research information about archaeological surveys and to exchange data about background radiation, or use satellite data to explore the varying power output from solar panels as they move in and out of shadow. An extended practical project provides students with an opportunity to explore one area in depth and it is likely that most projects will involve students making a link between content and applications.

The visit

As part of their coursework assessment in the AS year of the course, all students take part in an out-of-school visit to a place where they can see physics in use. Students are assessed on a short written report of their visit, in which they are expected to identify and explain the physics principles that they have observed in use and to recognise the purpose for which they are being used. The visit (at least in this guise) is unique to the Salters Horners course. It is envisaged that it will provide yet another way of helping students connect physics principles and real-life situations. In many cases, there might also be links between the visit and students' individual practical projects or work experience.

Salters Horners Physics Units

The Sound of Music – musical instruments and CD players

Working in Space – satellite power supply

Higher, Faster, Stronger – physics of sport

Good Enough to Eat – production, testing and packaging of food

Digging Up the Past – archaeology and geophysical surveying

Spare Part Surgery – hip replacement, corrective lenses and ultrasonic scanning

Transport on Track – rail transport, safety and control

The Medium is the Message – TV tubes, LCDs and fibre optics

Probing the Heart of the Matter – creation and detection of energetic particles

Reach for the Stars – formation and evolution of stars

Build or Bust – earthquakes, vibrations and building design

- Course materials include an article by a young space engineer describing his work, and an account of a theoretical physicist hitting on the idea of cosmological inflation

- It is envisaged that the visit will provide yet another way of helping students connect physics principles and real-life situations

Further information

Salters Horners Advanced Physics AS Level Books 1 and 2, Heinemann Educational, 1998.
GCE Salters Horners Physics AS/A Level Syllabus (pilot), Edexcel Foundation, London, 1998.

Course structure

As well as drawing the student into the study of physics, the contexts also provide a structure to each course unit which can differ from that in conventional courses.

A context-led unit is likely to involve more that one area of physics, and to require only a partial development of each area. In other units, students meet the same areas of physics in different contexts and develop their knowledge and understanding further. Students thus revisit some key areas several times and see their applicability to a wide range of situations, and interrelationships between different content areas are made explicit as they impinge on particular contexts.

For example, ideas about waves and superposition are explored in the context of standing waves in musical instruments and in the optical decoding of information from CDs, and a study of earthquake-resistant buildings links the physics of vibrations with mechanical properties of materials.

Evaluation

All the features of Salters Horners Physics described above are, to an extent, experimental. While they have been used successfully in other Salters courses, their use in an Advanced Physics course is novel. Twenty of the pilot centres have therefore been asked to provide the project team with detailed feedback from students, teachers and technicians. So far (September 1998), students and teachers have been asked about their initial expectations of the course.

Initial student perceptions

When students were asked (in a free-response question) what they expected to be the main advantages of the Salters Horners course, responses included the following and many more that were similar:

- It approaches physics in a manner which makes it more interesting, rather than just the concepts.

- Modern day applications makes it interesting.
- Puts subject into context
- To see how physics is used in the world
- Much more exciting and more life-based
- It includes a section on physics to do with buildings which is good for my career
- It teaches physics in terms of everyday life
- It will help me keep my career options open
- It is a new modern course and has close references to modern life

So far, then, the omens for a context-led physics course are good. The project team will continue to solicit feedback over the next two years. We will then be able to evaluate in much greater depth the extent to which various aspects of the context-led approach appeal to students and help them to 'make physics connect'.

DISCUSSION POINTS

- How does a teacher ensure that students really learn some physics from a visit?

- Has Salters Chemistry affected the female to male ratio of students?

- Compared to a conventional text and course, Salters requires that students read more.

14 Problems and Challenges in Higher Education

Ben Rudden

"They cannot transpose simple equations! They have never heard of Kirchhoff's laws! They don't know anything! They can't do derivations!

I am sure that during recent years nearly all of us teaching undergraduate physics will have experienced sentiments of that type. Perhaps it came as a surprise to discover that the O level had gone and that A level was not the same as that of the 'golden years' of the fifties and sixties! And did we respond with incredulity to the fact that the majority of 16-year-olds had never even heard the word 'physics'?

Of course, had we kept ourselves more informed about the major changes in the nature of the preparation of 16-plus students, particularly in physics and mathematics, we might have been better placed to respond to the rapid changes in schools and colleges, whereby the students entering our courses were bringing different skills and a lower knowledge-base than before.

The number of A level physics candidates continues to decline

On the other hand, we had lived with the 'whither physics' syndrome for decades. And in higher education, we had become adept at devising strategies to address the problem. The former polytechnics – which recruited on average, students with lower A level grades or students who had followed BTEC routes – encouraged by the CNNA, devised a variety of highly successful applied physics courses, usually with a sandwich component and very often of a modular structure. The old universities devised new final year modules very often related to cosmology in order to attract more students. In the meantime, the numbers taking A Level physics both absolutely, and relative to the age cohort, had continued to decline.

The eventual recognition that students entering the traditional three-year highly mathematical physics degree were incapable of absorbing the volume of material that was expected of them, culminated in 1993 in the four-year MPhys course. Students now have more time to absorb less material and thereby achieve a significantly fuller understanding of the subject. Moreover, time is available for the inclusion of broadening studies such as aspects of management or foreign languages.

However, only about 30 per cent of those who enter physics degrees eventually proceed to the MPhys or MSci qualification. It is from these that the required modest number of professional physicists will be drawn, though perhaps the majority will proceed to a variety of other careers, mainly non-physics related. Of course the other 50 per cent of the original input will aim for BSc (Honours) degrees in physics, and one has to ask what career opportunities will be available to them.

But there are more serious threats in terms of recruitment. Although, the actual numbers of students studying degree-level physics increased from 2821 in 1991 to 3183 in 1997, the proportion of undergraduates choosing physics continues to decline and now constitutes less than one percent of the total number of students.

Physics applications were down by two and a half per cent in 1998 although, according to an Institute of Physics survey, undergraduate courses are recruiting to target overall. But this does mean that departments which do fall significantly below target are vulnerable, particularly those which support modular schemes. We are all aware that a number of 'small' departments have recently closed or merged with other departments.

There is little current demand for physicists

Frankly, it has to be admitted that there is little current demand for physicists. Only about a half of those who graduated in physics in 1997 went into full-time jobs – not necessarily physics related – a third continued their studies for a higher degree, and the remainder became self-employed, part-time employees, went overseas, or are unemployed.

Indeed, the world as a whole is not crying out for physicists,

- In the four-year MPhys course students now have more time to absorb less material and gain a significantly fuller understanding of the subject

- It is little wonder that students are not attracted to physics research laboratory life which can only offer poor working conditions, poor pay, and short-term contracts

as the Australian and American experiences show. This is in sharp contrast with the employability of information technology, computing and other graduates (including physicists) with appropriate skills, the shortage of which has caused a significant increase in demand and, more relevantly, in expected starting salaries. In comparison it is little wonder that students are not attracted to physics research laboratory life which can only offer poor working conditions, poor pay, and short-term contracts.

Coupled with this, there is also the well recognised shortage of physics teachers and indeed it has been estimated that the shortfall now equals half the annual output of physics graduates. Consequently there is a serious lack of highly qualified teachers in school and college science departments to inspire students to contnue their study of physics.

We must therefore accept that despite many laudible initiatives, innovative schemes and even propaganda, physics still presents a negative image to the general public and particularly to young people and their parents. So where do we go now? The challenges seem insurmountable – all threats, and seemingly very few opportunities.

Post-16 education offers both challenges and opportunities
However, perhaps the most significant challenge and at the same time opportunity will present itself with the imminent fundamental changes in post-16 education, whereby A levels will be broadened. Students will be offered the chance to take five subjects in the first year and three in their second. Moreover, students will also have the opportunity to combine academic and vocational (GNVQ) units of study, which could ultimately lead to a unitised qualification system.

The Institute of Physics post-16 Initiative recognises that the coming changes provide the opportunity for a complete redesign of the 16–19 physics provision, not only in terms of content but also in approach, and a key element in this is the 'Making Physics Connect' theme. For the year 2002, universities will receive applications from potential students who wish to continue their study of physics after following the

OCR Advancing Physics AS/A2 course, and it is likely that that they will seek degree-level courses which continue the essential philosophy. Will such courses exist by then?

Time to ask ouselves some serious questions
The time has come, therefore, for those of us engaged in post-18 physics education to face up to the challenges and ask ourselves some serious questions

- Should we consolidate our efforts to encourage a small number of centres of excellence for the production of graduates who will become professional physicists?
- Should the remaining physics departments concentrate on research and post-graduate provision, and provide service physics for the dwindling number of other courses which require some physics input?
- Should there be a rolling programme of departmental closures, with staff transferred, similar to that involving East Anglia and Bath?
- Should we change the nature of our courses to attract students who will not necessarily become professional physicists, but whose skills are highly valued in a whole range of employment situations?
- Should we also seriously consider incorporating the 'Making Physics Connect' philosophy into our degree courses in the light of the coming changes to post-16 education?

In the early seventies, some of us may remember the SISCON project which was an attempt to provide a social, economic and political context for students studying science at A level, and also the 'Science and Society' degree courses which were offered at a number of universities. Excellent though they were, these courses eventually perished due to low recruitment.

Summary
A quarter of a century later perhaps we have the opportunity to resurrect some aspects of that movement, but indeed

- The shortfall of physics teachers has been estimated to now equal the annual output of physics graduates

- Here is a real chance to produce well-qualified physics graduates with a human face, equipped to respond to the challenges of the next century

achieve substantially more. Here is a real chance to produce well-qualified physics graduates with a human face, equipped to respond to the challenges of the next century.

We were caught unawares by the consequences of the National Curriculum. Let us make sure it does not happen again because the very survival of physics could depend on seizing this opportunity.

DISCUSSION POINTS

- Physics is expensive to teach. If a majority of those trained in physics go on to other careers, are there not cheaper ways to educate them?

- Is physics dying simply because it is an elite subject and HE is going through a transition from educating an elite to mass education?

- Providing graduates with prospects for employment in Europe will become more important. In many European countries, UK three-year degrees are regarded as inferior.

- The evidence is that new employment opportunities for physicists, such as work in the City or the development of quantum technologies, arise because employers need the high level physics knowledge of graduates, not their general skills

- The detachment of HE from physics education 14-19 continues to be a major problem.

- There may be opportunities for expansion of physics education in Access courses, or improving regional and lifelong learning opportunities for people.

A Selected Bibliography

There is a wealth of well-written and stimulating books to choose from, so my list is no more than indicative. Some of these are written for students, some for teachers, but most are for a general audience.

You might use this list to:

- add to a library in the physics lab
- suggest titles for a school or college library
- enhance teaching
- recommend extracurricular reading to students

Nature of science/physics

Solomon, J. (ed), *The Nature of Science*, ASE series, Heineman, London, 1989. Intended as student readers, these A5 booklets included three titles of particular interest to physics: *The Big Squeeze* (early ideas about the atmosphere), *Stars and Forces* (astronomy – the Greeks, Galileo, Newton and into the twentieth century), *Benjamin Franklin* (early ideas about electricity).

Dixon, B. and Holister G., *Ideas of Science*, Blackwell, Oxford, 1984. Intended for students, this book looks at some of the ideas and principles which guide scientists' work; for example, cause and effect, reproducibility, modelling, statistics. It also discusses the nature of experiments, the role of imagination in scientific work, and science–society issues.

Ellis, P., *What is Science?*, Pretext Publishing, 1990, and Ellis, P., *Science Changes*, Pretext Publishing, 1992. These two books were written as part of a modular science course, the first for Key Stage 3 (AT17 Nature of Science) and the second for Key Stage 4 (AT1 Scientific Investigation). Now available from Brecon Cottage, 33 Newbury St., Wantage, Oxon OX12 8BU.

Feynman, R., *The Character of Physical Law*, Penguin, London, 1992. Based on a lecture series, Feynman begins by elaborating gravitation as an example of physical law, then goes on to discuss the conservation laws and symmetry in physical law. He explains the different kind of theorising needed to explain behaviour at microscopic level before concluding with a discussion of how physical laws develop over time.

Krauss, L., *Fear of Physics*, Vintage, London, 1994. Another stimulating discussion about physical ideas and how physicists invent new ones.

Medawar, P., *The Limits of Science*, Oxford Paperbacks, 1984. A collection of three essays by a masterful writer, describing the 'great and glorious enterprise' that is science.

Wolpert, L., *The Unnatural Nature of Science*, Faber, London, 1992. Written for the layperson, this book tries to understand why science is so poorly understood by non-scientists, what science is and what it is not.

Science, technology & society

ASE, *SATIS 16-19*, Units 1-100
ASE 1983, *Science in Society* series.

Collins, H., & Pinch, T., *The Golem*, Cambridge University Press, 1993. Through a series of case studies, the authors aim to debunk science as a purely rational process and to show consensus is negotiated when experimental results are ambiguous.

Hales, M., *Science or Society: The Politics of the Work of Scientists*, Free Association Books, London, 1982.

Levidov, L., *Science as Politics*, Free Association Books, London, 1986.

History of science

Bernal, J. D., *The Extension of Man*, Paladin, St Albans, 1973. Based on a lecture series given at Birkbeck College, this is a history of physics to the end of the nineteenth century.

Bronowski, J., *The Ascent of Man*, Macdonald Futura, London, 1973. Derived from the BBC-TV series, with marvellous insights and turns of phrase, the story of science and its social context.

Gamow, G., *The Great Physicists from Galileo to Einstein*, Dover, New York, 1961. A narrative account encompassing both lives and works of historical figures, right up to the 1950s.

Wertheim, M., *Pythagoras' Trousers: God, Physics and the Gender Wars*, Fourth Estate, London, 1997. A social history of physics which argues that the priestly and masculine culture of physics has been a barrier to women.

Jungt, R., *Brighter than a Thousand Suns*, Penguin, London, 1960; and Rhodes, R., *The Making of the Atomic Bomb*, Simon & Schuster, New York, 1986. Two gripping accounts of the development of weapons which changed the world and sharpened the consciences of many physicists.

Hazen, R. M., *Superconductivity: The Break-through*, Unwin Hyman, London, 1988, A first-hand account of the scientific race to develop and explain superconductors at liquid nitrogen temperatures.

Beyond the Visible: 100 years of X-rays; Henri Becquerel and the Discovery of Radioactivity and *One Hundred Years of the Electron*, ASE. Three A4 student activity booklets.

Careers

Braben, D., *To Be a Scientist*, Oxford University Press, 1994. A contemporary view of 'how one becomes a scientist, what scientists do, and why what they do and how they do it is important for everyone'.

Wolpert, L., and Richards, A., *A Passion for Science*, Oxford University Press, 1988; *Passionate Minds*, Oxford University Press, Oxford, 1997. Two collections of interviews with prominent scientists, broadcast on BBC Radio 4. They portray vividly how scientists go about their work: where ideas come from, what role is played by chance and imagination. Their stated aim: to counter popular notions of scientists as cold and unfeeling technicians.

Biographies and autobiographies

Feynman, R., *Surely You're Joking, Mr Feynman*, Vintage, London, 1992; and *What Do You Care What Other People Think?*, Bantam, New York, 1989. Two collections of anecdotes, giving at once both humorous and serious insight into scientific thinking. Students find him shocking and attractive.

Regis, E., *Who Got Einstein's Office?*, Penguin, London, 1987. A portrait of the Institute for Advanced Study and many who worked there, described as a 'cracking good read . . . the flavour of the place, the people, and their subjects shines through'.

Hart-Davis, A., and Bader, P., *The Local Heroes Book of British Ingenuity*, Sutton Publishing, Stroud, 1997. This book of the popular TV series tells the stories of over fifty pioneers of science and technology from around the country.

Wali, K. C., *Chandra: a biography of S. Chandrasekhar*, University of Chicago Press, 1991. There are of course biographies of Einstein, Feynman, Faraday, Heisenberg, Hoyle, Newton, Rutherford, Schrodinger and many others.

Cultural links

Science has inspired some of the century's most imaginative writing.
Drama:
Brecht , B., *The Life of Galileo*, Methuen, London, 1963; Frayn, M., *Copenhagen*, Methuen, London, 1998; Stoppard, T., *Hapgood*, Samuel French, London, 1988, and *Arcadia*, Faber, London, 1997.
Fiction:
Calvino, I., *Cosmicomics*, Abacus, London, 1969; Gratzer, W., (ed) *The Longman Literary Companion to Science*, Longman, London, 1989.

Miller, A., *Insights of Genius*, Copernicus/ Springer-Verlag, New York, 1996. A study of links between science and the visual arts, focussing on the early twentieth century.

Collections of essays and edited extracts from science writing

Covering a wide range of topics, edited extracts can introduce you to writers whose full work you then want to read. Because each is short, these are excellent for reading in short bursts.
Carey, J., (ed) *The Faber Book of Science*, 1995.
Dixon, B., (ed) *From Creation to Chaos*, Blackwell, Oxford, 1989.
Ferris, T., (ed) *The World Treasury of Physics, Astronomy & Maths*, Little Brown, New York, 1991.

Gardner, M., (ed), *Science: Good, Bad and Bogus*, Oxford University Press, 1981; *The Sacred Beetle*, Oxford University Press, 1985; and *The New Age*, Prometheus Books, Buffalo, 1991.
Perutz ,M., *Is Science Necessary? Essays on Science & Scientists*, Barrie and Jenkins, London, 1989.
Von Baeyer, H. C., *The Fermi Solution*, Random House, New York, 1993,

Multiculturalism and Race Equality

Race Equality and Science Teaching Handbook, ASE, 1994. The best single source on this theme, it is both thorough and comprehensive and includes references to many other sources.

Reiss, M.J., *Science Education for a Pluralist Society*, Open University Press, 1993. The author argues that a pluralist society requires greater equality of standing between science as carried out and perceived by different cultural, ethnic, gender, class, ability and religious groups. He offers suggestions of how specific topics might be taught.

Gill, D., and Levidow, L. (ed), *Anti-Racist Science Teaching*, Free Association Books, London, 1987. Using case studies, this book shows that science and technology embody distinctive values and cultural assumptions, including racist ones, and that these permeate science teaching. It proposes a theory and practice of anti-racist education through science.

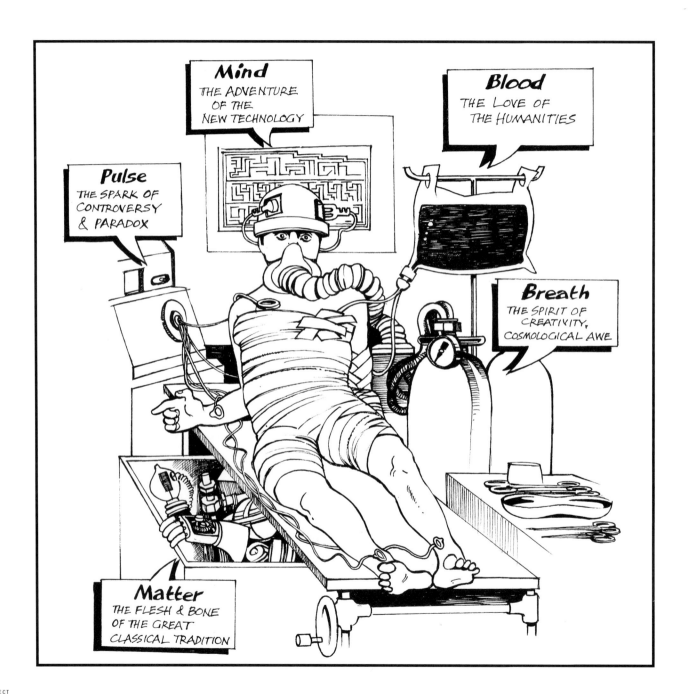

Loose Ends

Derived from a conference held in October 1998

Enlivening versus essential

Throughout the booklet, the argument about connections in physics (with the past, with culture, with lives) shifts from their enlivening lessons to their being essential to a proper education in physics. It really is important to work out what is essential. But that may not mean that certain cases become essential – that everyone should hear about Faraday, for example. That makes no sense, and does not use the teacher's own knowledge and enthusiasm well. What needs to be essential must be certain *kinds* of idea or connection, not the particular cases one person would choose.

Making best use of teaching time?

If the purpose of going 'off syllabus' is to enliven the learning, we can afford to use precious lesson time for this. It may provide the 'activation energy' needed for better understanding and harder work. Shortcuts then become possible elsewhere.

Accessibility of courses

As larger numbers of people stay on at school post-16 and return to study as adults, it is important that we have on offer physics courses which are accessible to a wider audience. For them, the connections may form a vital bridge which opens the subject up. In this sense, all professional physicists shouldbe concerned with the public image of the subject: exclusive or inclusive?

Broadening the curriculum

We need to get away from thinking primarily of two year A level courses. New AS courses, such as the AQA course 'Science for Public Understanding', would fit well into a broadened Year 12, as would mixing academic with vocational units. But will the government provide the additional funding to make possible increased teaching hours for students?

Teachers' interests

Contributors are at their most convincing when offering something which has personally enthused them. A mistake easily made is to suppose that these can be passed on to others as they are, and that the enthusiasm and commitment will travel well. It must be true of teachers, as of contributors, that they too are at their most convincing when offering something which has personally enthused them. The question is not how to provide teachers with set pieces, but how to help them, with their limited preparation time, find what they want.

Ordinary lives

Perhaps this booklet is weakest on making connections to ordinary lives. What does a technician in a fully automated factory actually do? What is the patient's experience of a high technology body scan?

Qualifications and Curriculum Authority

The QCA not only permits A level physics courses to connect, but expects them to do so:

> *A and AS specifications in physics should encourage students to: appreciate how physics has developed and is used in present day society; and show the importance of physics as a human endeavour which interacts with social, philosophical, economic and industrial matters.*

It cannot be healthy that the normal response of physics examiners to this is at best an embarrassed smile.

Websites

The technology of the web offers real possibilities for providing and updating material for teachers to choose and use. And for them to discuss how it worked.

Making Physics Connect – Some Ways Forward

1 For The Teacher

The teacher has a large measure of control over styles of teaching and learning. One of the most important things a physics teacher can do is to allow personal qualities to have a place in this abstract subject – not only glimpsing science personalities outside the classroom, but also acknowledging the personal in the classroom. We recommend a variety of activities in the classroom, so that students experience groupwork, discussion and debate, as well as the usual range of practical work and lectures. But teachers also need to develop a personal style: choose ways of doing things so that students experience variety but you remain confident.

A large part of 'Making Physics Connect' is contextualising the physics content; however, a balance must be struck so that content remains the main focus of classroom activity. Remember you can encourage particular students to further pursue contextual interests outside the classroom.

To varying degrees, course texts address the need for context. Going beyond these, you can draw on teachers in other departments and on standard library searches. Librarians in schools and colleges are usually very pleased to assist with curriculum development, provided you are clear enough about what you want. Many are now adept at drawing on the resources of the Internet. We hope the Institute of Physics website will develop so that teachers can exchange good ideas through a Bulletin Board, and resources for 'Making Physics Connect' will be posted for all to use. It is important that teachers have opportunities for in-service training, and that it includes these themes.

2 For Exam Boards

The QCA insists that all physics syllabuses should aim to include 'social, philosophical, economic and industrial matters'. This can be taken seriously, provided the syllabus allows teaching time for making physics connect. Examiners appreciate that there is considerable scope for exploring industrial applications of physics. Philosophical issues abound when considering the nature of science. What are the limits of science, particularly in relation to big questions such as 'Who are we?', 'Where are we?' and 'What can we know?'. The human drama in science which we would like students to encounter in classrooms very often has a social, not to say a moral, dimension. Depending on the outlook of teacher and students, even spiritual questions may arise from moral issues too; for example, the many illustrations that physics drives weapons technologies raises questions about human purposes.

What is a more difficult problem for Boards is deciding whether and how such learning can be assessed. Unless it is assessed, at least in part, it will continue to be treated as optional, and many teachers will avoid it altogether. In developing the Advancing Physics AS/A2 course we are exploring how contextual learning might be assessed: i) as part of a written paper, and ii) through coursework. It will be useful for physics subject panels to share experience in this area.

3 For Universities

In order to maintain student numbers, universities need to concern themselves with the image of physics. A physics education is relevant not just to those who will use it directly in their daily working lives: modelling skills and general learning skills are increasingly valued by employers.

Developing new courses for regional and lifelong learning may in effect mean looking again at 'physics for poets' courses. 'Making Physics Connect' themes could also improve the delivery of physics undergraduate courses. An ongoing dialogue between pre-university and university staff would help both groups.

4 For Teacher Training

It is vital for the future of physics that high quality new entrants are attracted to the teaching profession. Present school physics courses are largely based on pre-twentieth century ideas. Courses which are more up-to-date in outlook and conceptual content, and which better employ graduates' university knowledge, may prove more attractive to potential teacher recruits. And graduates who are motivated to work with young people are likely to be interested also in humanising science. With all the current regulation of initial teacher training courses, previous good practice of including 'Making Physics Connect' themes should not be squeezed out and forgotten.